ACTIVATION ANALYSIS
WITH CHARGED PARTICLES

ELLIS HORWOOD SERIES IN ANALYTICAL CHEMISTRY

Series Editors: Dr R. A. CHALMERS and Dr MARY MASSON, University of Aberdeen

Consultant Editor: Prof. J. N. MILLER, Loughborough University of Technology

S. Allenmark	Chromatographic Enantioseparation — Methods and Applications
G.E. Baiulescu & V.V. Coşofreţ	Application of Ion Selective Membrane Electrodes in Organic Analysis
G.E. Baiulescu, P. Dumitrescu & P.Gh. Zugravescu	Sampling
G.E. Baiulescu, C. Patroescu & R.A. Chalmers	Education and Teaching in Analytical Chemistry
G.I. Bekov & V.S. Letokhov	Laser Resonant Photoionization Spectroscopy for Trace Analysis
S. Bance	Handbook of Practical Organic Microanalysis
H. Barańska, A. Łabudzińska & J. Terpiński	Laser Raman Spectrometry
K. Beyermann	Organic Trace Analysis
O. Budevsky	Foundations of Chemical Analysis
J. Buffle	Complexation Reactions in Aquatic Systems: An Analytical Approach
D.T. Burns, A. Townshend & A.G. Catchpole	Inorganic Reaction Chemistry Volume 1: Systematic Chemical Separation
D.T. Burns, A. Townshend & A.H. Carter	Inorganic Reaction Chemistry: Volume 2: Reactions of the Elements and their Compounds: Part A: Alkali Metals to Nitrogen, Part B: Osmium to Zirconium
E. Casassas & S. Aligret	Solvent Extraction
S. Caroli	Improved Hollow Cathode Lamps for Atomic Spectroscopy
J. Churáček	New Trends in the Theory & Instrumentation of Selected Analytical Methods
E. Constantin, A. Schnell & M. Mruzek	Mass Spectrometry
R. Czoch & A. Francik	Instrumental Effects in Homodyne Electron Paramagnetic Resonance Spectrometers
T.E. Edmonds	Interfacing Analytical Instrumentation with Microcomputers
J.K. Foreman & P.B. Stockwell	Automatic Chemical Analysis
Z. Galus	Fundamentals of Electrochemical Analysis, Second Edition
J. Gasparič & J. Churáček	Laboratory Handbook of Paper and Thin Layer Chromatography
S. Görög	Steroid Analysis in the Pharmaceutical Industry
T. S. Harrison	Handbook of Analytical Control of Iron and Steel Production
J.P. Hart	Electroanalysis of Biologically Important Compounds
T.F. Hartley	Computerized Quality Control: Programs for the Analytical Laboratory
Saad S.M. Hassan	Organic Analysis using Atomic Absorption Spectrometry
M.H. Ho	Analytical Methods in Forensic Chemistry
Z. Holzbecher, L. Diviš, M. Král, L. Šůcha & F. Vláčil	Handbook of Organic Reagents in Inorganic Chemistry
A. Hulanicki	Reactions of Acids and Bases in Analytical Chemistry
David Huskins	Electrical and Magnetic Methods in On-line Process Analysis
David Huskins	General Handbook of On-line Process Analysers
David Huskins	Optical Methods in On-line Process Analysis
David Huskins	Quality Measuring Instruments in On-line Process Analysis
J. Inczédy	Analytical Applications of Complex Equilibria
Z.K. Jelínek	Particle Size Analysis
M. Kaljurand & E. Küllik	Computerized Multiple Input Chromatography
R. Kalvoda	Operational Amplifiers in Chemical Instrumentation
I. Kerese	Methods of Protein Analysis
S. Kotrlý & L. Šůcha	Handbook of Chemical Equilibria in Analytical Chemistry
J. Kragten	Atlas of Metal-ligand Equilibria in Aqueous Solution
A.M. Krstulović	Quantitative Analysis of Catecholamines and Related Compounds
F.J. Krug & E.A.G. Zagotto	Flow Injection Analysis in Agriculture & Environmental Science
V. Linek, V. Vacek, J. Sinkule & P. Beneš	Measurement of Oxygen by Membrane-covered Probes
C. Liteanu & S. Gocan	Gradient Liquid Chromatography
C. Liteanu, E. Hopîrtean & R. A. Chalmers	Titrimetric Analytical Chemistry
C. Liteanu & I. Rîcă	Statistical Theory and Methodology of Trace Analysis
Z. Marczenko	Separation and Spectrophotometric Determination of Elements
M. Meloun, J. Havel & E. Högfeldt	Computation of Solution Equilibria
M. Meloun, J. Militky & M. Forina	Chemometrics in Instrumental Analysis: Solved Problems for IBM PC
O. Mikeš	Laboratory Handbook of Chromatographic and Allied Methods
J.C. Miller & J.N. Miller	Statistics for Analytical Chemistry, Second Edition
J.N. Miller	Fluorescence Spectroscopy
J.N. Miller	Modern Analytical Chemistry
J. Minczewski, J. Chwastowska & R. Dybczyński	Separation and Preconcentration Methods in Inorganic Trace Analysis
T.T. Orlovsky	Chromatographic Adsorption Analysis
D. Pérez-Bendito & M. Silva	Kinetic Methods in Analytical Chemistry
B. Ravindranath	Principles and Practice of Chromotography
V. Sediveč & J. Flek	Handbook of Analysis of Organic Solvents
R.M. Smith	Derivatization for High Pressure Liquid Chromatography
R.V. Smith	Handbook of Biopharmaceutic Analysis
K.R. Spurny	Physical and Chemical Characterization of Individual Airborne Particles
K. Štulík & V. Pacáková	Electroanalytical Measurements in Flowing Liquids
O. Shpigun & Yu. A. Zolotov	Ion Chromatography in Water Analysis
J. Tölgyessy & E.H. Klehr	Nuclear Environmental Chemical Analysis
J. Tölgyessy & M. Kyrš	Radioanalytical Chemistry, Volumes I & II
J. Urbanski, et al.	Handbook of Analysis of Synthetic Polymers and Plastics
M. Valcárcel & M.D. Luque de Castro	Flow-Injection Analysis: Principles and Applications
C. Vandecasteele	Activation Analysis with Charged Particles
J. Veselý, D. Weiss & K. Štulík	Analysis with Ion-Selective Electrodes
F. Vydra, K. Štulík & E. Juláková	Electrochemical Stripping Analysis
N. G. West	Practical Environment Analysis using x-ray Fluorescence Spectrometry
F.K. Zimmermann & R. E. Taylor-Mayer	Mutagenicity Testing in Environmental Pollution Control
J. Zupan	Computer-supported Spectroscopic Databases
J. Zýka	Instrumentation in Analytical Chemistry

ACTIVATION ANALYSIS WITH CHARGED PARTICLES

C. VANDECASTEELE
Research Director
Belgian National Fund for Scientific Research, and
Laboratory for Analytical Chemistry
Institute for Nuclear Sciences
Ghent University, Ghent, Belgium

Editor:
R. A. CHALMERS
Department of Chemistry
University of Aberdeen

ELLIS HORWOOD LIMITED
Publishers · Chichester

Halsted Press: a division of
JOHN WILEY & SONS
New York · Chichester · Brisbane · Toronto

First published in 1988 by
ELLIS HORWOOD LIMITED
Market Cross House, Cooper Street,
Chichester, West Sussex, PO19 1EB, England
The publisher's colophon is reproduced from James Gillison's drawing of the ancient Market Cross, Chichester.

Distributors:

Australia and New Zealand:
JACARANDA WILEY LIMITED
GPO Box 859, Brisbane, Queensland 4001, Australia

Canada:
JOHN WILEY & SONS CANADA LIMITED
22 Worcester Road, Rexdale, Ontario, Canada

Europe and Africa:
JOHN WILEY & SONS LIMITED
Baffins Lane, Chichester, West Sussex, England

North and South America and the rest of the world:
Halsted Press: a division of
JOHN WILEY & SONS
605 Third Avenue, New York, NY 10158, USA

South-East Asia
JOHN WILEY & SONS (SEA) PTE LIMITED
37 Jalan Pemimpin # 05–04
Block B, Union Industrial Building, Singapore 2057

Indian Subcontinent
WILEY EASTERN LIMITED
4835/24 Ansari Road
Daryaganj, New Delhi 110002, India

© 1988 C. Vandecasteele/Ellis Horwood Limited

British Library Cataloguing in Publication Data
Vandecasteele, C. (Carlo), *1949–*
Activation analysis with charged particles.
1. Activation analysis
I. Title
543'.0882

Library of Congress Card No. 88–23289

ISBN 0–7458–0175–7 (Ellis Horwood Limited)
ISBN 0–470–21204–7 (Halsted Press)

Typeset in Times by Ellis Horwood Limited
Printed in Great Britain by Hartnolls, Bodmin

Table of contents

1

Introduction

Activation analysis is an analytical method that uses the production of radionuclides from the analyte element(s) in the sample for the determination of this (these) element(s). The sample is irradiated with particles that undergo nuclear reactions with the atomic nuclei of the element(s) of interest, and the radionuclides produced are identified and determined by measurement of the radiation emitted upon disintegration.

The first artificial radionuclide was produced by Joliot and Curie in 1934 [1], and von Hevesy and Levi first applied artificial radioactivity to analytical chemistry in 1936 [2]. Systematic development of activation analysis, however, only began around 1950. At the time, the first nuclear research reactors had begun operation and suitable particle accelerators had become available. Activation analysis then underwent rapid development: the development of neutron activation analysis (NAA) by use of a nuclear reactor was most rapid and spectacular, as can, for instance, be seen in the proceedings of the 'Modern Trends in Activation Analysis' Conferences, which, starting from 1961, took place every 4–5 years. Reactor neutron activation analysis has been developed into a reliable and powerful analytical method, for trace element analysis, allowing the determination of over 60 elements, with good accuracy and low detection limits. Some important advantages are as follows.

1. Only minimal sample handling and treatment before the irradiation are required, thus minimizing risk of contamination and loss of trace elements. Contamination and trace element loss are indeed the most important sources of error when trace elements are determined by other sensitive analytical methods, for instance atomic-absorption spectrometry, ICP–atomic-emission spectrometry and ICP–mass spectrometry, which all require dissolution of the sample before the measurement.
2. In many cases a purely instrumental determination, i.e. without chemical separations, is feasible.
3. When a chemical separation is required, carrier can be added to avoid problems

that occur when such separations are performed at very low concentrations, and also to allow easy determination of the chemical yield. Moreover, the method is free from reagent blanks.

4. Calibration is relatively easy: the pure element or a chemical compound with well known composition and purity can be used as a standard.

An important limitation is that some elements cannot be determined by reactor neutron activation analysis, for instance the light elements (hydrogen, beryllium, boron, carbon, nitrogen and oxygen), lead, etc. Moreover, for some matrices the activity induced in the matrix is high and long-lived, so chemical separations are required to separate the radionuclide(s) of interest from the matrix activity. These separations must sometimes be done in hot-cells for reasons of radiation protection. Neutron activation analysis has been treated in detail in several books, for instance [3–6].

Activation analysis with charged particles (CPAA) also already has a relatively long history. The first CPAA experiment was conducted by Livingood and Seaborg in 1938. It concerned the determination of traces of gallium in iron by deuteron activation. As long ago as the first Modern Trends in Activation Analysis Conference, Albert [7] summarized the application of CPAA for the determination of carbon, nitrogen and oxygen in metals. Ricci and Hahn [8] described in 1965 a convenient method for standardization, whereby the activity in the sample is compared to that of a standard irradiated separately with particles of the same energy. This standardization method is often used even today.

The development of CPAA was, however, much slower than that of NAA for several reasons.

1. CPAA is in general more complex than NAA, because several nuclear reactions occur simultaneously. For instance, in irradiation with 20-MeV protons the following nuclear reactions occur: (p, n), (p, pn), (p, d), (p, 2n), (p, 2p), (p, α), p, t) . . . . Nuclear interferences are therefore also more frequent than in NAA with thermal neutrons, where mainly (n, γ) reactions occur.
2. Charged particles have, in contrast to neutrons and photons, a limited penetration range in matter. Since in addition the stopping power and the range are matrix-dependent, standardization is more complicated than in NAA. Moreover, accurate data for the stopping power and the range in all elements have only recently been compiled in a readily accessible form.
3. Since charged particles lose their energy in a limited thickness of the sample, the sample heats up during the irradiation, which may cause problems with several kinds of samples.
4. Particle accelerators providing charged particles with suitable energy were less often available for this type of research than were nuclear reactors.

Light charged particles — protons, deuterons, tritons, helium-3, helium-4 — are most frequently used for activation analysis. The particle energies range from a few MeV to about 40 MeV. More recently heavy ions have also been used in activation analysis.

The analytical results given in some early papers were rather imprecise and of unknown accuracy, so the method was sometimes even considered as only semi-quantitative. During the past 10–15 years, however, several sources of error have been studied in detail.

1. For the determination of light elements, such as carbon, nitrogen and oxygen, in bulk or massive samples, e.g. metals, the influence of activation of the element at the surface of the sample was studied. It was shown that post-irradiation removal of a surface layer containing the activity formed at the surface is essential for reliable results to be obtained.
2. The standardization methods were studied in detail and methods were developed that do not introduce approximations. In addition, use was made of more reliable stopping power and range data.
3. In some cases an instrumental analysis is feasible, in other cases, the radionuclide of interest must be separated chemically from interfering radionuclides. The chemical separations often require a high decontamination factor and their development requires a great deal of effort.

Because of these studies, the accuracy and precision obtainable with CPAA have improved significantly. The accuracy has been demonstrated by various inter-laboratory and inter-method comparisons.

Owing to its inherent complexity, CPAA is not used frequently as a routine analytical method, as NAA often is. The method is, however, very valuable as a complement to NAA for special applications.

1. CPAA may be considered as 'the' reference method for the determination of low concentrations of light elements in metals and semiconductors.
2. It can also be used for the determination of traces of medium and heavy elements in metals at concentration levels below the detection limit of most 'classical' chemical methods, a particular advantage of CPAA for these applicaitons being its freedom from reagent blanks. CPAA allows instrumental analysis of some matrices that are difficult to analyse by NAA because of high and long-lived induced activity, and some elements that are impossible or very difficult to determine by NAA can be determined by CPAA.
3. CPAA is useful for analysing 'standards' for use in less 'absolute' analytical techniques, such as X-ray fluorescence spectrometry (XRF).
4. CPAA is often valuable as an independent 'absolute' method for the certification of reference materials.

In principle, CPAA is suitable for the analysis of almost every matrix. The major portion of the literature deals, however, with metals and semiconductor materials. These are indeed easy to irradiate and surface contamination can easily be removed, e.g. by chemical etching after the irradiation. The analysis of environmental, geological, and biological materials has also been studied.

Charged particles have also often been used as projectiles in prompt activation analysis. The γ-rays from de-excitation of nuclides, scattered particles or charged particles from e.g. (d, p) or (d, α) reactions are then measured during the irradiation, by means of germanium semiconductor or silicon surface-barrier detectors. These methods are particularly suitable for depth profiling and determination of surface contamination, and are treated in detail in [9] and [10]. The present book does not cover these applications, but is limited to 'bulk' analysis by charged particle activation analysis.

REFERENCES

[1] F. Joliot and I. Curie, *Nature*, 1934, **133**, 201.
[2] G. von Hevesy and H. Levi, *Kgl. Danske Videnskab. Selskab. Math.-fys. Medd.*, 1936, **14**, 5.
[3] D. De Soete, R. Gijbels and J. Hoste, *Neutron Activation Analysis*, Wiley-Interscience, London, 1972.
[4] P. J. Elving, V. Krivan and I. M. Kolthoff (ed.), *Treatise on Analytical Chemistry*, Part I, 2nd Ed., Vol. 14, Wiley, New York, 1986.
[5] S. Amiel, *Non-destructive Activation Analysis with Nuclear Reactors and Radioactive Neutron Sources*, Elsevier, Amsterdam, 1981.
[6] J. Tölgyessy and I. M. Kyrš, *Nuclear Analytical Chemistry*, Horwood, Chichester, in the press.
[7] P. Albert, in *Proceedings of the 1961 International Conference on Modern Trends in Activation Analysis*, College Station, Texas, 15–16 December, 1961, M. W. Brown (ed.), Activation Analysis Research Laboratory, Texas, 1961.
[8] E. Ricci and R. L. Hahn, *Anal. Chem.*, 1965, **37**, 742.
[9] G. Deconninck, *Introduction to Radioanalytical Physics*, Elsevier, Amsterdam, 1978.
[10] D. Brune, B. Foreman and B. Persson, *Nuclear Analytical Chemistry*, Studentlitteratur, Lund, 1984.

2

Theory

This chapter treats some theoretical aspects of CPAA, such as nuclear reactions with charged particles, stopping power and range, and standardization. Theoretical aspects that apply to NAA as well as to CPAA, e.g. nuclear decay, are not treated, but reference is made to some books on activation analysis [1–3].

2.1 NUCLEAR REACTIONS WITH CHARGED PARTICLES

Nuclear reactions with charged particles may vary from simple, e.g. (p, n), to quite complex, e.g. (p, ^{3}He 2n), depending mainly on the nature and the energy of the projectile and on the target nucleus. Usually, several nuclear reactions occur simultaneously. The complexity of the situation is illustrated by Fig. 2.1, which can

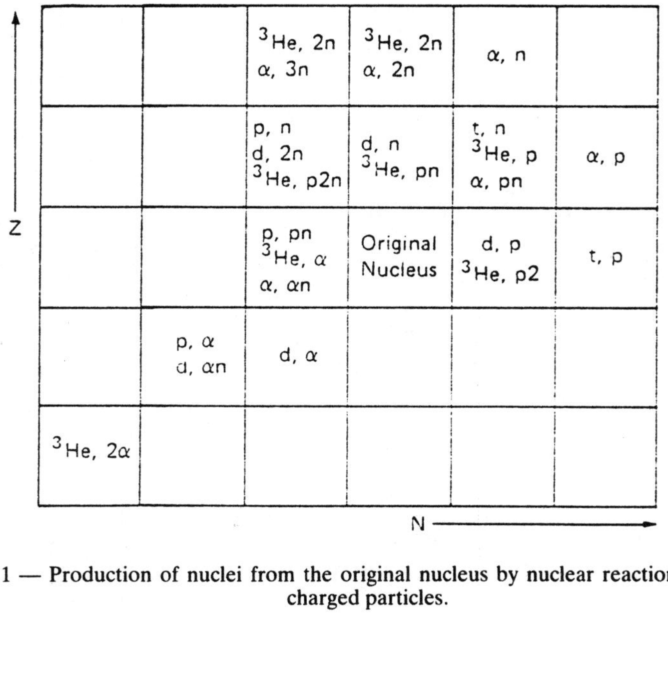

		^{3}He, 2n α, 3n	^{3}He, 2n α, 2n	α, n	
		p, n d, 2n ^{3}He, p2n	d, n ^{3}He, pn	t, n ^{3}He, p α, pn	α, p
Z		p, pn ^{3}He, α α, αn	Original Nucleus	d, p ^{3}He, p2	t, p
	p, α d, αn	d, α			
^{3}He, 2α					

N ⟶

Fig. 2.1 — Production of nuclei from the original nucleus by nuclear reactions induced by charged particles.

be considered as an overlay of the Chart of the Nuclides [4] and indicates the nuclei produced from a given original nucleus by proton (p), deuteron (d), triton (t), helium-3 (^{3}He) or helium-4 (α) induced nuclear reactions.

2.1.1 Q-Value and threshold energy

The Q-value of a nuclear reaction is the energy liberated per reaction event. For the nuclear reaction A(a, b)B, Q is given by Eq. (2.1)

$$Q = (m'_A + m'_a - m'_b - m'_B)c^2 \tag{2.1}$$

where m'_A, m'_a, m_b, m'_B = masses of the neutral atoms of the nuclides A, a, b and B; c = velocity of light.

When the masses are expressed in atomic mass units (amu), Q (expressed in MeV) is given by Eq. (2.2):

$$Q = 931.50(m'_A + m'_a - m'_b - m'_B) \tag{2.2}$$

The compilation by Keller *et al.* [5] gives Q-values for a large number of nuclear reactions.

When Q is positive, the reaction is called exoergic, when it is negative, the reaction is called endoergic. For an endoergic reaction the threshold energy, i.e. the minimum energy needed by a projectile to make the reaction energetically possible, is given by Eq. (2.3).

$$E_T = -\frac{m_A + m_a}{m_A}Q \approx -\frac{A_A + A_a}{A_A}Q \tag{2.3}$$

where m_A, m_a = masses of target nuclide A and projectile a; A_A, A_a = mass numbers of A and a.

Equation (2.3) is a consequence of the law of conservation of momentum.

2.1.2 Cross-section and excitation function

The probability of a nuclear reaction is expressed by means of the cross-section $\sigma(E)$, which has the dimensions of a surface, with a value that depends on the energy. A unit often used is the barn (b), which is equal to 10^{-24} cm^2.

Consider a beam of i particles with an energy E, incident per second on an infinitesimally thin layer of target material with density ρ and thickness dl. The target material contains n_A nuclides A per gram. The number of nuclear reaction events A(a, b)B per second is then equal to $n_A i\sigma(E)\rho$dl. The curve that gives the cross-section as a function of energy is called the excitation function.

2.1.3 Potential barrier and Coulomb threshold energy

When a positively charged particle approaches a nucleus, it is repelled by the electrostatic field. At a distance r from the nucleus the potential energy is equal to

$$\frac{kZ_A Z_a e^2}{r}$$

where $k = 8.99 \times 10^9 \, \text{N.m}^2.\text{C}^{-2}$; Z_A, $Z_a =$ the atomic numbers of the target nucleus and the projectile; $e =$ charge of an electron, $1.602 \times 10^{-19} \, \text{C}$.

When the projectile and the target nucleus are in contact, the potential energy is given by Eq. (2.4).

$$B = k\frac{Z_A Z_a e^2}{R_A + R_a} \tag{2.4}$$

where R_A, $R_a =$ nuclear radii of the target nucleus and projectile. B is called the Coulomb potential barrier.

If it is taken into account that during the collision the compound nucleus is not immobile, but acquires a kinetic energy determined by the law of conservation of momentum, the Coulomb threshold energy (MeV) is approximately given by Eq. (2.5)

$$E_C = 1.02\frac{(A_A + A_a)Z_A Z_a}{A_A(A_A^{1/3} + A_a^{1/3})} \tag{2.5}$$

Figure 2.2 gives the Coulomb threshold energy calculated by means of Eq. (2.5) for protons, deuterons, helium-3 and helium-4 as a function of the atomic number of the target nucleus.

If a nuclear reaction is energetically possible, it can take place with high probability only if the energy of the incident particle exceeds E_C. Charged particles with an energy below the Coulomb threshold energy can only yield nuclear reactions by 'tunnelling' through the barrier. The cross-section then decreases rapidly with the particle energy.

When the incident particle has an energy in excess of the Coulomb threshold energy, it can penetrate into the nucleus and undergo the influence of the nuclear forces. In the nucleus there exists an overall potential consisting of large negative terms of nuclear origin and (for a charged particle) of a Coulomb term. Figure 2.3 shows schematically this overall potential for a neutron and a photon. The nuclear potential barrier, i.e. the maximum potential energy, is zero for neutrons and positive for charged particles. In order to release a neutron from a given energy level in the potential well, it is sufficient to supply the separation or binding energy, S_n, i.e. the difference in energy for the particle at infinity and in the nucleus. To release a proton the energy required is $S_p + B$. As soon as the proton has left the nucleus, B is released as kinetic energy.

2.1.4   Mechanism of nuclear reactions; shape of excitation functions

The exact mechanism of nuclear reactions is not yet fully understood, but there exist semi-empirical models giving a simplified picture.

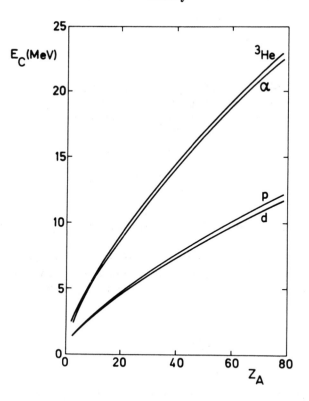

Fig. 2.2 — Coulomb threshold energy E_C for protons (p), deuterons (d), helium-3 (^{3}He) and helium-4 (α) as a function of the atomic number Z_A of the target nucleus.

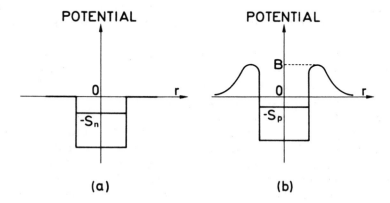

Fig. 2.3 — Nuclear potential barrier for a neutron (a) and a proton (b).

The first model was proposed by Bohr in 1936 and assumes that a nuclear reaction takes place in 2 steps:

(1) capture of the incident particle with formation of a compound nucleus C:

$$A + a \rightarrow C \tag{2.6}$$

(2) decay of the compound nucleus with release of b, a particle (n, p, d, α . . .) or a γ-photon:

$$C \rightarrow B + b \tag{2.7}$$

The compound nucleus has an excitation energy U_C given by Eq. (2.8):

$$U_C = \frac{m_A}{m_A + m_a} E_a + S_a \tag{2.8}$$

where E_a = kinetic energy of the particle a in the laboratory system; S_a = binding energy of a in the compound nucleus.

The excitation energy is rapidly randomly distributed over the nucleons in the compound nucleus. The compound nucleus exists in a quasi-stationary state: it can decay by particle emission, but its lifetime is long ($10^{-19} - 10^{-14}$ sec) compared to the time required for a particle to traverse the nucleus ($10^{-23} - 10^{-20}$ sec). None of the nucleons has sufficient energy to escape immediately, but because of statistical fluctuations of the energy distribution, at a given time sufficient energy, at least equal to the separation energy, concentrates on a nucleon (or a cluster of nucleons) and allows it to escape. For a charged particle, extra energy is required to overcome the nuclear potential barrier. The most probable fluctuations are those where only a fraction of the excitation energy is concentrated on the escaping particle, so that it acquires a lower energy than the maximum possible energy and the remaining nucleus is still in an excited state. If the original excitation energy is high enough, sequential emission of several particles may thus occur, each with a relatively low energy.

The cross-section σ_{ab} of the nuclear reaction A(a, b)B can be written as:

$$\sigma_{ab} = \sigma_C(a) P_b \tag{2.9}$$

where $\sigma_C(a)$ = cross-section for the formation of the compound nucleus; P_b = probability that C decays with emission of b.

It can be shown [6, 7] that for a charged particle with an energy ξ in excess of the Coulomb potential barrier B, $\sigma_C(a)$ is approximately given by Eq. (2.10)

$$\sigma_C(a) = \pi(R_A + R_a + \lambda)^2 \left(1 - \frac{B}{\xi}\right) \tag{2.10}$$

where R_A, R_a = radii of the target nucleus and projectile; $\lambda = h/2\pi p$, with h = Planck's constant; p = momentum at the centre of the mass system, given by

$$p = \frac{m_A m_a}{m_A + m_a} v_a = \mu v_a \tag{2.11}$$

where v_a = velocity of the incident particle in the laboratory system; μ = reduced mass; B = Coulomb potential barrier, given by Eq. (2.4); ξ = total kinetic energy in the centre of the mass system:

$$\xi = \frac{m_A}{m_A + m_a} E_a = \frac{1}{2} \mu v_a^2 \tag{2.12}$$

where E_a = kinetic energy of the incident particle in the laboratory system.

It follows from Eq. (2.10) that $\sigma_C(a)$ approaches $\pi(R_a + R_A)^2$ if ξ is much higher than B. Sometimes B is replaced by $k_f B$, k_f being a factor less than unity, in order to get better agreement with experimental results.

The compound nucleus has discrete energy levels, called virtual levels, since they can decay by particle emission, in contrast to bound levels, which can only decay by γ-photon emission. Each level has a mean lifetime τ and a width Γ, related to τ by the uncertainty principle:

$$\Gamma = \frac{h}{2\pi\tau} \tag{2.13}$$

where h is Planck's constant. Each decay mode corresponds to a partial width Γ_x, the total width for a given level being given by the sum of the partial widths:

$$\Gamma = \Gamma_\gamma + \Gamma_p + \Gamma_\alpha + \Gamma_n + \ldots \tag{2.14}$$

If the levels are sufficiently well separated, they can be individually excited with particles of the correct energy. The cross-section of the reaction considered is high when the excitation energy corresponds exactly with an excited state of the compound nucleus. The excitation function then shows resonance peaks. For light nuclei and low excitation energy the energy levels of the compound nucleus are often widely spaced and pronounced resonance peaks occur in the excitation function. Below 10 MeV the most common reactions are of the (p, γ), (p, α), and (p, n) type. The Coulomb potential barrier makes nuclear reactions between heavy nuclei and low-energy charged particles impossible.

When the energy of the incident particle a and thus the excitation energy increases, the width of the energy levels increases, because more decay modes are possible, and the distance between the levels decreases. Eventually, the levels overlap to yield a continuum and it is no longer possible to excite only one level, so the excitation function no longer shows distinct resonances. Depending on the energy of the emitted particles, several particles may be emitted to carry all the excitation energy away. The total cross-section is thus divided between reactions of the (α, n), $(\alpha, 2n)$, (α, pn), $(\alpha, 3n)$ ... type.

Figure 2.4 gives the excitation functions for the (α, n), $(\alpha, 2n)$ and $(\alpha, 3n)$

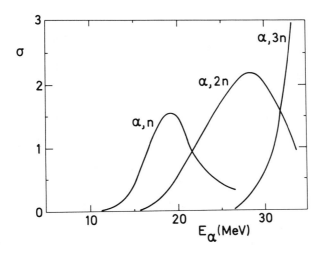

Fig. 2.4 — Excitation functions for the (α, n), $(\alpha, 2n)$ and $(\alpha, 3n)$ reactions of ^{115}In. σ is given in arbitrary units. (Reproduced by permission, from G. M. Temmer, *Phys. Rev.*, 1949, **76**, 424. Copyright 1949, American Physical Society.)

reactions of ^{115}In. At low energy only the (α, n) reaction occurs ($Q = -7.2$ MeV, $E_T = 7.5$ MeV). Between 11 and 18 MeV the cross-section increases rapidly, because the probability for penetration of the Coulomb potential barrier ($E_C = 16.0$ MeV) increases. At about 16 MeV the excitation energy is sufficient to allow the emission of 2 neutrons and the $(\alpha, 2n)$ reaction ($Q = -14.6$ MeV; $E_T = 15.1$ MeV) starts to occur in competition with the (α, n) reaction. The (α, n) reaction goes through a maximum at ca. 19 MeV and then decreases in favour of the $(\alpha, 2n)$ reaction. Above 26 MeV the compound nucleus sometimes still retains, after emission of 2 neutrons, sufficient energy to emit a third neutron. The $(\alpha, 3n)$ reaction ($Q = -24.3$ MeV, $E_T = 25.2$ MeV) starts to occur at the expense of the $(\alpha, 2n)$ reaction. The shape of these excitation functions is a consequence of the fact that each particle emitted carries away only a fraction of the available excitation energy. The competition between energetically possible nuclear reactions is governed by the relative probability for emission of different particles, e.g. neutrons, protons, deuterons, tritons, helium-3 and helium-4 from the compound nucleus. The emission probability for a particle b depends on the available energy $U_C - S_b$ and on the nuclear potential barrier, the barrier suppressing emission of charged particles, mainly for heavy nuclei.

When nuclei with an atomic number around 30 are irradiated with 5–15 MeV protons, mainly (p,n) and (p,α) reactions occur. When the energy increases to 15–25 MeV, reactions of the (p, 2n), (p, pn), (p, 2p), and (p, αn) type start to predominate. (p, ³He), (p, t) and (p, d) reactions may also occur, albeit with a lower cross-section and with features that can better be explained from the 'pick-up' mechanism (see below). When the energy increases further, reactions with emission of 3 or more particles, such as (p, 3n), (p, p2n), (p, α2n), and (p, αpn) dominate. Figure 2.5 gives excitation functions for some proton-induced reactions of ⁶³Cu.

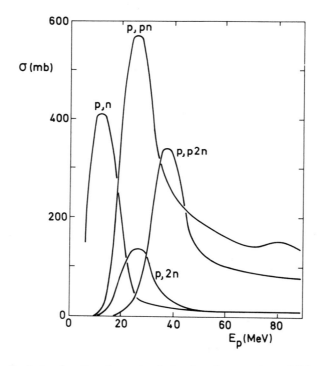

Fig. 2.5 — Excitation functions for proton-induced nuclear reactions of ⁶³Cu. (Adapted by permission, from J. W. Meadows, *Phys. Rev.*, 1953, **91**, 885. Copyright 1953, American Physical Society.)

When the incident particle is an α-particle instead of a proton, the reaction pattern is roughly similar (Fig. 2.4). ³He and ³H particles are also emitted as a consequence of 'stripping' (see below). Reactions with deuterons and helium-3 (Fig. 2.6) also show roughly the same pattern, but with more important contributions from 'stripping' reactions.

When the atomic number of the target increases, the increasing nuclear potential barrier progressively suppresses emission of charged particles. For bismuth only (p, xn) reactions occur, the number of neutrons emitted, x, amounting to 4 or 5 at

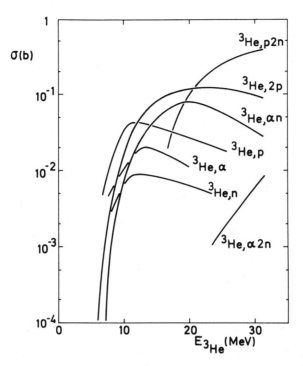

Fig. 2.6 — Excitation functions for ^{3}He-induced nuclear reactions of ^{63}Cu [10].

around 50 MeV. The maximum cross-section of a charged particle induced nuclear reaction usually ranges from 10 mb to 1 b.

The direct interaction model differs from the compound nucleus model in that it does not assume that the energy of the incident particle is divided between the nucleons of the target nucleus. It is assumed that the incident particle collides with one or a few nucleons in the target nucleus, one or more nucleons being directly ejected. The reaction thus does not take place through the formation of a compound nucleus and it can be expected that the kinetic energy of the emitted particles is higher. Excitation functions of some nuclear reactions indeed show a tail up to relatively high energy (40–80 MeV), whereas the compound nucleus model predicts a steep decrease when more complex nuclear reactions become possible. Reactions whereby only a part of the incident composite particle (e.g. a deuteron) interacts with the nucleus are also classified as direct interactions. The part that does not interact continues its path after having been deflected. Such reactions were first observed for deuterons and are called 'stripping reactions'. They are of the (d, p) and (d, n) type. Other examples are (^{3}He, d) and (α, d). The reverse of 'stripping' is the so-called 'pick-up' process, whereby a composite particle (deuteron, helium-3) is formed by interaction of an incident particle (proton) with a nucleon or a group of nucleons in the nucleus.

At higher energies spallation reactions occur; essentially any spallation reaction that is energetically possible occurs. However, in general, the products in the immediate neighbourhood of the target element are found in the highest yields.

Experimentally determined excitation functions of nuclear reactions have been compiled by Keller et al. [10] and by McGowan and Milner [11]. More recent data are given in *Nuclear Data Tables* and in *Atomic Data and Nuclear Data Tables*.

2.2 STOPPING POWER AND RANGE

On their path through matter, charged particles (protons, deuterons, helium-3, helium-4 and heavier ions) lose energy mainly by interaction with electrons. This eventually leads to excitation or ionization of atoms and molecules or to dissociation of molecules. Charged particles have a range: when a mono-energetic beam of charged particles loses energy on its path through matter, the number of particles in the beam does not change, and eventually all the particles are stopped after roughly the same path length. The range is the mean path length for a large number of particles. Because of the large mass of the incident particle (m_a) relative to the electron mass (m_e) the fractional energy loss per collision is small, at most $4m_e/m_a$. A large number of collisions is thus required to stop each particle, so the effect of fluctuations of the energy loss per collision is minimal. Furthermore, the deviation during each collision is small. The path is therefore, to a good approximation, linear and the range is close to the so-called projected range, i.e. the projection of the path on the original direction of motion.

The stopping of a charged particle in matter occurs in 3 phases:

(1) at a sufficiently high velocity the particle is stripped of all its electrons and energy loss occurs by interaction with the electrons of the target material (electronic stopping);
(2) at velocities comparable with the velocity of the electrons in its *K*-shell, the particle starts to pick up electrons from the target material, so that the mean charge decreases gradually from Z_a to $Z_a - 1$; the mechanism of energy loss is still the same provided that $Z_a - 1$ is greater than zero;
(3) at velocities comparable to those of the valence electrons of the target material, energy loss occurs mainly through elastic collisions between the particle and the atoms of the target material (nuclear stopping).

2.2.1 Definitions and theory
The stopping power $S(E)$ of a material for a particle with energy E is a function of the energy and is defined by Eq. (2.15)

$$S(E) = \frac{-dE}{dx}(E) ; \qquad dx = \rho dl \qquad (2.15)$$

where ρ = density; dl =distance corresponding to the energy loss dE. The stopping power can be expressed in $J.m^2.kg^{-1}$. As a practical unit $MeV.cm^2.g^{-1}$ is generally used.

At high energy (>1 MeV) the stopping power is well described by the Bethe equation [12]:

$$S(E) = \frac{4\pi e^4 Z_a^2 N_A Z_A}{m_e v_a^2 M} \left[\ln \frac{2m_e v_a^2}{I} + \ln \frac{1}{1-\beta^2} - \beta^2 - \frac{C}{Z_A} \right] \qquad (2.16)$$

where e = charge of the electron; Z_a = atomic number of the projectile; N_a = Avogadro's constant; Z_A = atomic number of the target material; m_e = electron mass; v_a = velocity of the projectile; M = molecular weight of the target material; β = velocity of the projectile divided by the velocity of light, I = effective ionization potential.

The effective ionization potential I is the main parameter in the theory. Theoretical calculations led to Bloch's rule [13], according to which I is proportional to Z_A, the proportionality constant being ca. 10 eV. Experimental data show that the proportionality constant decreases as Z_A increases. More complicated equations for I have been proposed [14, 15]. The term C/Z_A, representing the shell-corrections, takes into account the fact that some electrons do not participate in the slowing-down process. The correction for a given shell i asymptotically approaches zero for $v_a \gg v_i$, where v_i is the orbital velocity of a bound electron. Since for heavy elements the electrons in the inner shells have relativistic velocities, even at high energy the latter condition ($v_a \gg v_i$) is not fulfilled. Theoretical methods of calculation for the shell corrections have been proposed [15–18].

Equation (2.16) shows that the stopping power for a given particle is, to a first approximation, inversely proportional to the particle energy. For particles with the same velocity v_a, the stopping power is proportional to the square of the atomic number of the projectile Z_a^2. Therefore, for protons of energy E, deuterons of energy $2E$, and tritons of energy $3E$, which all have the same velocity, the stopping power is the same. A ^{4}He-particle of energy $4E$ has the same velocity as a proton of energy E, but, since its atomic number is twice that of the proton, the stopping power is 4 times greater for the ^{4}He-particle.

The range of a particle with energy E_I is defined by Eq. (2.17):

$$R(E_I) = \int_0^{E_I} \frac{dE}{S(E)} \qquad (2.17)$$

and corresponds to the mean path length for particles of energy E_I. $R(E_I)$ is often expressed in g/cm^2. Sometimes use is made of the projected range, corresponding to the mean distance travelled by the particle in the original direction. The projected range is somewhat smaller than the range, since the path is not completely straight [12]. Andersen and Ziegler [19] give curves that show the range and the projected range as a function of the energy. For light elements (e.g. carbon) the curves coincide above ca. 300 keV; for heavy elements (e.g. tantalum) they coincide above ca. 2 MeV.

The stopping power given by Eq. (2.16) is only an average value. The number of collisions with electrons per unit path length, as well as the energy loss per collision, is subject to statistical fluctuations, so the energy loss may differ for particles with the same energy that traverse the same distance in the target. This is called 'straggling'. An incident mono-energetic beam of particles with energy E_I will, after traversing a distance in the target, show an energy distribution, which has been calculated by Vavilov [20]. The distribution of the energy loss is asymmetrical with a tail towards high-energy losses. When the average energy loss Δ_0 increases, the distribution approaches closer and closer to a Gaussian distribution. On condition that [21]:

$$x \geqslant 3 \times 10^{-4} \frac{A_A E_I^2}{Z_A A_a^2} \tag{2.18}$$

where $x = \rho l$ with l = distance traversed by the particle (cm); ρ = density of the target material (g/cm³); A_A, Z_A = mass number and atomic number of the target; A_a = mass number of the projectile; E_I = incident energy (MeV), the distribution of the energy loss Δ is given by Eq. (2.19)

$$f(x, \Delta) = \frac{1}{1.06 \Gamma_0} \exp\left[-\frac{(\Delta - \Delta_0)^2}{0.36 \Gamma_0^2}\right] \tag{2.19}$$

Γ_0, the full width at half maximum of the Gaussian distribution, expressed in MeV, is given by Eq. (2.20):

$$\Gamma_0 = 0.93 Z_a \left(\frac{Z_A x}{A_A}\right)^{1/2} \tag{2.20}$$

Δ_0 can be deduced from Eq. (2.21):

$$R(E_I - \Delta_0) = R(E_I) - x \tag{2.21}$$

Equation (2.19) is due to Bohr [22] and is in good agreement with experimental results.

For a mono-energetic proton beam with $E_I = 10$ MeV, the energy distribution of the particles after a path length of 0.1182 g/cm² in aluminium can be calculated as follows, the parameters used being $Z_A = 13$, $A_A = 27$, $Z_a = 1$, $A_a = 1$, $E_I = 10$. The approximation to a Gaussian distribution is valid if

$$x > 0.062 \text{ g/cm}^2 \tag{2.22}$$

The average energy loss Δ_0 is given by:

$$R(10 - \Delta_0) = R(10) - 0.1182 \tag{2.23}$$

This yields:

$$\Delta_0 = 5 \, \text{MeV} \tag{2.24}$$

The full width at half maximum, Γ_0, is given by:

$$\Gamma_0 = 0.93 \left(\frac{13 \times 0.1182}{27} \right)^{1/2} = 0.22 \, \text{MeV} \tag{2.25}$$

For a proton beam degraded from 10 to 5 MeV, Γ_0 is 0.20, 0.22, 0.25 and 0.28 MeV for carbon, aluminium, zirconium and tungsten, respectively. For 20 MeV helium-4, degraded to 10 MeV, the values are 0.22, 0.24, 0.28 and 0.32 MeV, respectively. Straggling thus differs only slightly for different materials, as shown experimentally by Tran and Tousset [23].

Because of straggling, particles from an originally mono-energetic beam do not all have the same range, but the distribution of the ranges is approximately a Gaussian distribution around $R(E_I)$ with a variance that can be calculated according to [24].

2.2.2 Compilations

The most complete and recent compilations of stopping power data are those of Andersen and Ziegler [19] for hydrogen ions and of Ziegler [25] for helium ions. For high energies these authors use the Bethe equation (2.16) as the theoretical basis. Since in this equation the velocity of the projectile appears, but not the mass, all absolute experimental data for the stopping power for protons, deuterons, and tritons were considered as a function of E', the energy expressed in keV/amu. Starting from I-values from the literature [26–28] the shell correction term for every experimental point was calculated by means of the Bethe equation. Through the points that gave, for a given element, the shell correction term as a function of E', a smooth curve was drawn, with a shape based on theoretical considerations [26, 29]. Shell correction curves for hydrogen, helium, lithium, beryllium, boron, carbon, nitrogen, oxygen, neon, aluminium, argon, calcium, nickel, copper, selenium, krypton, zirconium, molybdenum, silver, tin, xenon, gadolinium, tantalum, platinum, gold, lead, and uranium were thus obtained. In order to obtain C/Z_A values smoothly varying as a function of Z_A, some I-values and the shape of some shell-correction curves had to be altered somewhat. For 27 elements the power series of the form

$$C/Z_A = a_8 + a_9 \ln E' + a_{10}(\ln E')^2 + a_{11}(\ln E')^3 + a_{12}(\ln E')^4$$

$$= \sum_{i=0}^{4} a_{i+8} (\ln E')^i \tag{2.26}$$

that best fits the shell-correction curve, was calculated. For the other elements, where insufficient experimental data were available, the shell-correction term was obtained by interpolation of Z_A, and a similar best fitting power series was calculated. For low energies (<1 MeV) the equations of Lindhard and Scharff [30] and of Varelas and Biersack [31] were taken as the theoretical basis. Since the energy range below 1 MeV is of little significance for CPAA, we will not go into the details of the procedure used.

The compilation by Andersen and Ziegler [19] gives equations that allow the stopping power (expressed in 10^{-15} eV.cm^2.atom^{-1}) to be calculated as a function of the energy expressed in keV/amu, by using the compiled cofficients a_2–a_{12}. To calculate the stopping power, expressed in MeV.cm^2.g^{-1}, the equations given in Table 2.1 can be used, Eq. (2.31) being a modified form of Eqs. (2.16) and (2.26).

Table 2.1 — Equations allowing the calculation of the stopping power (MeV.cm^2.g^{-1}) as a function of the energy (MeV) for protons and deuterons.

$$E' = 10^3 E/m_H \tag{2.27}$$

If $10 \leqslant E' < 1000$

$$ST_L = a_2(E')^{0.45} \tag{2.28}$$

$$ST_H = \frac{a_3}{E'} \ln\left(1 + \frac{a_4}{E'} + a_5 E'\right) \tag{2.29}$$

$$\frac{1}{ST} = \frac{1}{ST_L} + \frac{1}{ST_H} \tag{2.30}$$

If $1000 < E' < 100000$

$$ST = \frac{a_6}{\beta^2}\left[\ln\frac{a_7\beta^2}{1-\beta^2} - \beta^2 - \sum_{i=0}^{4} a_{i+8}(\ln E')^i\right] \tag{2.31}$$

$$\beta^2 = 1 - \left[\frac{931.5 m_H}{E + 931.5 m_H}\right]^2 \tag{2.32}$$

$$S(E) = \frac{N_A}{10^{21} M} ST \tag{2.33}$$

E = energy of the incident particle (MeV); m_H = mass of hydrogen nucleus, for the proton: 1.00728 amu, for the deuteron: 2.01355 amu; a_j = coefficients compiled by Andersen and Ziegler [19] (j = 2, 3, . . . 12); $S(E)$ = stopping power (MeV.cm^2.g^{-1}); N_A = Avogadro's constant, 6.02204×10^{23}/mole; M = molecular weight of the element considered (g/mole).

From a comparison between experimental and calculated results, Andersen and Ziegler [19] conclude that their equations are, at high energy, accurate to within 1%, and, at 600 keV per nucleon, to within 3%. The ratios of the stopping power of a given element to that of aluminium at the same energy, obtained by precise measurements and by calculation were also compared. If each calculated stopping power was considered to have an uncertainty of 0.5%, no systematic deviations were observed. Table 2.2 gives the equations for helium-3 and helium-4.

Table 2.2 — Equations allowing the calculation of the stopping power ($MeV.cm^2.g^{-1}$) as a function of the energy (MeV) for helium-3 and helium-4.

If $0.001 < E < 10$

$$ST_L = a_1(10^3 E)^{a_2} \qquad\qquad (2.34)$$

$$ST_H = \frac{a_3}{E}\ln(1 + a_4/E + a_5 E) \qquad\qquad (2.35)$$

$$\frac{1}{ST} = \frac{1}{ST_L} + \frac{1}{ST_H} \qquad\qquad (2.36)$$

If $10 < E < 300$

$$(EE) = \ln(1/E) \qquad\qquad (2.37)$$

$$ST = \exp[a_6 + a_7(EE) + a_8(EE)^2 + a_9(EE)^3] \qquad\qquad (2.38)$$

$$S(E) = \frac{N_A}{10^{21} M} ST \qquad\qquad (2.39)$$

For helium-3, $E = 4/3 E_{3_{He}}$
E = energy of the incident ^{4}He-particle, or 4/3 of the energy of the incident ^{3}He-particle (MeV); a_j = coefficients compiled by Ziegler [25] (j = 1, 2, ... 9); $S(E)$ = stopping power ($MeV.cm^2.g^{-1}$); N_A = Avogadro's constant; M = molar mass of the element considered (g/mole).

As an illustration, Table 2.3 gives the stopping power of nickel for protons of

Table 2.3 — Stopping power and range for protons in nickel.

Energy, MeV	Stopping power, $MeV.cm^2.g^{-1}$	Range, g/cm^2	Range, μm
1	128.4	0.0056	6.3
2	85.6	0.0154	17.3
5	46.9	0.0654	73.4
10	28.6	0.207	233
15	21.2	0.413	464
20	17.1	0.677	761
25	14.4	0.997	1120

different energies, calculated as shown in Table 2.1, and the ranges calculated by numerical integration using Eq. (2.17).

Table 2.4 gives the stopping power and the range for 10 and 20 MeV protons in

Table 2.4 — Stopping power (MeV.cm².g⁻¹) and range (g/cm²) for protons in different pure elements.

Z_A	Element	10 MeV		20 MeV	
		S	R	S	R
6	C	40.8	0.138	23.3	0.477
13	Al	33.9	0.170	19.7	0.573
28	Ni	28.6	0.207	17.1	0.677
48	Cd	22.5	0.270	13.7	0.860
82	Pb	17.8	0.353	11.1	1.089

different pure elements, and Table 2.5 gives the range for 10 MeV protons, deuterons, helium-3 and helium-4 in nickel.

2.2.3 Mixtures and compounds
For mixtures and compounds, the stopping power is calculated by means of the additivity rule of Bragg and Kleeman [32]:

$$S(E) = \sum_{i=1}^{n} \omega_i S_i(E) \qquad (2.40)$$

where $S(E)$ = stopping power of a mixture or a compound with n components, at energy E; $S_i(E)$ = stopping power of component i at energy E; ω_i = mass fraction of component i.

When the stopping power is known as a function of the energy, Eq. (2.17) can be used to calculate the range.

Sometimes, to calculate the range in a mixture or a compound, use is made of Eq. (2.41)

$$\frac{1}{R(E_I)} = \sum_{i=1}^{n} \frac{\omega_i}{R_i(E_I)} \qquad (2.41)$$

where $R(E_I)$ = range of a particle with energy E_I in a mixture or a compound with n components (g/cm²); $R_i(E_I)$ = range in component i.

For compounds, the additivity rule of Bragg and Kleeman is an approximation that does not take into account the influence of chemical bonding. The actual stopping power is lower than that calculated, since the outer electrons are more strongly bound in a compound and thus play a lesser role in the slowing-down of the incident particle [33]. It can be expected that the difference between the calculated and the experimental stopping power is negligible at high energy and increases as the energy decreases.

Several authors have experimentally investigated the validity of the additivity rule. Langley and Blewer [33] found that the stopping power of Er_2O_3 is 0–9% lower than expected for 2.5–0.5 MeV protons and 4–13% for 2.5–0.5 MeV helium-4. Ishii

Table 2.5 — Range for 10 MeV protons, deuterons, helium-3 and helium-4 in nickel.

Particle	R, g/cm^2
p	0.207
d	0.131
^{3}He	0.0260
^{4}He	0.0223

et al. [34] and Blondiaux *et al.* [35, 36] studied the influence of chemical bonding on the stopping power of aluminium oxide, titanium dioxide, zinc oxide, niobium pentoxide and tantalum pentoxide for 2.3–2.7 MeV protons and of beryllium oxide for 1.9–2.2 MeV helium-4. The ratio of the stopping power of the metal oxide to that of the metal was compared to the value expected from Eq. (2.40). For protons the values agreed within 1.1%, for 1.9–2.2 MeV helium-4 on beryllium oxide deviations of 2.1–8.8% occurred. Baglin and Ziegler [37] tested the additivity rule for 2 MeV helium-4 and a number of compounds. Within an experimental uncertainty of ca. 2%, the calculated and experimental values agreed for silicon dioxide, aluminium oxide, silicon nitride, aluminium nitride, tungsten nitride and silicon carbide. The additivity rule of Bragg and Kleeman is thus sufficiently accurate for application in activation analysis. The experimentally observed deviations are generally small and occur only at low energy, in general below the threshold energy or the Coulomb threshold energy of most nuclear reactions.

2.3 STANDARDIZATION

Generally in CPAA a thick sample and a thick standard, i.e. thicker than the range in the material considered, are irradiated separately with particles with an energy E_I. As explained in Section 2.2, charged particles have a limited range in matter and the stopping power and the range are matrix-dependent. If the sample and the standard are not identical, the ranges and hence the depths activated differ.

Because of the relatively low cross-section for nuclear reactions with charged particles and of the limited range, attenuation of the beam in the sample and the standard is usually negligible in the energy range considered. The beam intensity i (particles per second) for a beam with initial intensity i_0, after traverse of a distance l in a material is given by Eq. (2.42):

$$i = i_0 \exp[-\rho N_A \sigma l / M] \tag{2.42}$$

where ρ = density; N_A = Avogadro's constant; σ = total cross-section; M = molecular weight of the target element. It is assumed that the target element is mono-isotopic.

For 10 MeV protons in aluminium (ρ = 2.702 g/cm^3; M = 26.98 g/mole), assuming that σ is 10^{-24} cm^2, at the end of the range (0.170 g/cm^2) i amounts to 99.6% of i_0. At lower energy and for helium-3 and helium-4, attenuation is even less important, because of the smaller range.

When an infinitely thin sample is irradiated with charged particles of energy E, perpendicular to the surface, the increase in the number of radionuclides formed by the nuclear reaction considered, per unit of time, is given by

$$\frac{dN^*(t)}{dt} = n_A i\sigma(E)dx - \lambda N^*(t) \tag{2.43}$$

where $N^*(t)$ = number of product radionuclides at time t; n_A = number of nuclides per unit mass that give the nuclear reaction considered; i = beam intensity, number of particles incident on the target per unit time; $\sigma(E)$ = cross-section of the reaction at energy E; $dx = \rho dl$, *with* ρ = density, dl = target thickness; λ = decay constant of the radionuclide produced [given by Eq. (2.44)]

$$\lambda = (\ln 2)/t_{1/2} = 0.693/t_{1/2} \tag{2.44}$$

with $t_{1/2}$ = half-life.

The first term of Eq. (2.43) gives the number of radionuclides formed per unit time (see Section 2.1.2), and λN^* is the number of radionuclides that decay per unit time.

Integration of Eq. (2.43) gives the activity A in becquerels (disintegrations/sec) after an irradiation time t_b, on condition that i is constant during the irradiation:

$$A = \lambda N^*(t_b) = n_A i\sigma(E)dx(1 - e^{-\lambda t_b}) \tag{2.45}$$

The factor $1 - e^{-\lambda t_b}$ is called the saturation factor; for irradiation times very short compared to the half-life it is equal to λt_b, for an irradiation time equal to 1, 2, 3 half-lives it is equal to 1/2, 3/4, 7/8 . . . , and for irradiation times that are long compared to the half-life it approaches 1.

The cross-section is a function of the energy. When the sample is not infinitely thin, $\sigma(E)$ changes with depth in the sample, as the particle energy changes with depth. For a thick sample, the activity (thick target yield) is given by Eq. (2.46):

$$A = n_A i(1 - e^{-\lambda t_b}) \int_0^{R(E_I)} \sigma(x)dx = n_A i(1 - e^{-\lambda t_b}) \int_0^{E_I} \frac{\sigma(E)dE}{S(E)} \tag{2.46}$$

where $R(E_I)$ = range at E_I.

Since

$$n_A = \frac{10^{-6}cN_A\theta}{M} \tag{2.47}$$

where c = concentration of the analyte element (μg/g); θ = natural isotopic abundance of the nuclide that gives the nuclear reaction considered (as discussed in Section 3.5, natural abundances may not be constant for some elements), M = molar mass of the analyte element, the concentration of the analyte element is given by

$$c = \frac{A}{i(1-e^{-\lambda t_b})} \frac{1}{\int_0^{E_1} \frac{\sigma(E)dE}{S(E)}} \left(\frac{10^6 M}{N_A \theta}\right)$$ (2.48)

With Eq. (2.48), the concentration can be deduced from

— the activity A in becquerels (disintegrations/sec);
— the beam intensity i (particles/sec);
— the irradiation time t_b (sec);
— the excitation function $\sigma(E)$;
— the stopping power $S(E)$.

However, measurements of the absolute activity and of the absolute beam intensity are difficult and, moreover, (absolute) excitation functions of nuclear reactions induced by charged particles are not sufficiently well known to allow accurate analyses by use of Eq. (2.48). Therefore, a sample (X) and a standard (S) are usually irradiated separately with particles of the same energy. Provided that the isotopic abundance of the nuclide considered is the same for the sample and the standard:

$$c_X = c_S \frac{A_X}{i_X[1-e^{-\lambda t_{b,X}}]} \frac{i_S[1-e^{-\lambda t_{b,S}}]}{A_S} \frac{\int_0^{E_1} \frac{\sigma(E)dE}{S_S(E)}}{\int_0^{E_1} \frac{\sigma(E)dE}{S_X(E)}}$$ (2.49)

If the activity of the sample and the standard is measured with the same detection efficiency, the ratio of the absolute activities A_X/A_S can be replaced by the ratio of the measured count-rates (counts/sec) a_X/a_S: a is the count-rate at the end of the irradiation and is given by Eq. (2.50):

$$a = \frac{\lambda C e^{\lambda t_w}}{1-e^{-\lambda t_m}}$$ (2.50)

where C = number of counts during the measuring time t_m; t_w = waiting time between the end of the irradiation and the start of the measurement.
Equation (2.50) can be derived from Eq. (2.51)

$$C = \int_{t_w}^{(t_w+t_m)} a e^{-\lambda t} dt$$ (2.51)

In order to calculate the concentration of the analyte element by using Eq. (2.49), we must know:

— the concentration in the standard c_S;
— the ratio of the count-rates a_X/a_S;
— the ratio of the beam intensities i_S/i_X;
— the irradiation times $t_{b,X}$ and $t_{b,S}$;
— the correction factor $F(E_I)$, defined by Eq. (2.52).

$$F(E_I) = \frac{\int_0^{E_I} \dfrac{\sigma(E)dE}{S_S(E)}}{\int_0^{E} \dfrac{\sigma(E)dE}{S_X(E)}} \tag{2.52}$$

The correction factor $F(E_I)$ takes the different stopping powers of the sample and the standard into account. Since below the threshold energy E_T, $\sigma(E) = 0$, $F(E_I)$ can also be written as:

$$F(E_I) = \frac{\int_{E_T}^{E_I} \dfrac{\sigma(E)dE}{S_S(E)}}{\int_{E_T}^{E_I} \dfrac{\sigma(E)dE}{S_X(E)}} \tag{2.53}$$

The correction factor $F(E_I)$ can be calculated, if the excitation function of the reaction of interest, $\sigma(E)$, is known at least relatively, and the stopping powers of the standard $S_S(E)$ and the sample $S_X(E)$ are known. As discussed in Section 2.2, the second condition is fulfilled, but the relative excitation function is not known for all nuclear reactions of interest.

2.3.1 Numerical integration
By use of Eq. (2.53), $F(E_I)$ can be calculated by numerical integration, on condition that $\sigma(E)$ for the considered reaction and $S(E)$ for the sample and the standard are known. When this method is used frequently, it is best to develop a simple computer program to carry out the calculations.

2.3.2 Stopping power and accuracy
It appears from Eq. (2.49) that the stopping power data have a direct repercussion on the final analytical result. The accuracy of the stopping power data for pure elements was discussed in Section 2.2.2. According to Andersen and Ziegler [19], their data are accurate to within 1% at high energy and to within 3% at 600 keV per nucleon. For most nuclear reactions the cross-section below 1 MeV is only a small fraction of that at the maximum of the excitation function. Furthermore, the stopping power increases as the energy decreases: the ratio of the stopping power for 1 MeV protons

to that for 5 MeV protons is for instance 3.07 for aluminium and 2.32 for lead. The activity induced below 1 MeV is thus in general only a small fraction of the total activity, so the lower accuracy of the stopping power data at low energy plays only a minor role.

When the sample or the standard is a chemical compound, possible deviations from the additivity rule of Bragg and Kleeman must also be taken into account. Moreover, when the bulk composition of the sample is not well known, additional systematic errors are introduced.

2.3.3   Excitation function and accuracy

In order to determine the absolute cross-section of a nuclear reaction, a thin target with an accurately known thickness and composition is irradiated with a beam of charged particles with a given energy, the beam intensity being measured during the irradiation. The number of nuclear reaction events in the target is determined by measurement of the radionuclides produced or by detection of the particles emitted during the reaction. In the former case, care must be taken that the reaction products are not lost by recoil. Catcher foils are therefore placed before and after the target to stop recoil nuclei.

In order to determine the excitation function, the cross-section is measured at different energies. Use is also made of the stacked-foil technique, wherein a stack of target foils containing the element of interest (e.g. mica for nuclear reactions on oxygen) is irradiated. The total thickness of the foils must be sufficient to stop the beam and, in principle, the energy loss in each foil must be small, which may be difficult to achieve, especially at low energy. Often foils of another material are placed between the foils of the target material to degrade the energy, and the activity in the target foils is measured after the irradiation. In this way, the absolute excitation function can be determined, if the beam intensity is accurately measured and the detection efficiency of the detector is taken into account. Of course, determination of the relative excitation function presents fewer experimental difficulties.

When an excitation function is determined, in addition to random errors, e.g. due to counting statistics, systematic errors may occur from:

— inaccurate determination of the beam intensity;
— inaccurate determination of the efficiency of the detector;
— interfering nuclear reactions;
— inaccurate incident energy;
— inaccurate thickness and composition of the target foils.

When the stacked foil technique is used to determine a relative excitation function, the first two sources of error do not apply. With this method, inaccurate measurement of the thickness of the foils and inaccurate range-energy data may also result in an inaccurate incident energy for each foil.

Excitation functions experimentally determined by different authors often do not agree satisfactorily. In order to evaluate the influence of the excitation function data on the final analytical result, $F(E_I)$ can be calculated by numerical integration with different experimentally determined excitation functions. In the following sections only nuclear reactions used for the determination of light elements in metals will be considered. In this case the standard is usually a light element matrix (boron, graphite, quartz . . .) with an (average) atomic number lower than that of the sample. When heavy elements are determined, the standard is usually a heavy element matrix (e.g. lead metal or lead oxide for lead) with an atomic number higher than that of the sample, and some conclusions are opposite to those for the determination of light elements.

2.3.3.1   *Reaction* $^{12}C(d, n)^{13}N$

Figure 2.7 gives the excitation function for the $^{12}C(d, n)^{13}N$ reaction, experimentally determined by several authors. Below 5 MeV, curve 2, experimentally determined by Wilkinson [39] differs most from the others, probably because a stack of polyethylene foils was irradiated with 20 MeV deuterons, so that a small error in the incident energy, the thickness, or the ranges results in a large error in the position of the maximum of the excitation function. Table 2.6 gives $F(E_I)$ for 3 MeV deuterons, with the four excitation functions. The deviation increases with increasing atomic number of the sample but in all cases is below 1%, except for the excitation function of Wilkinson [39], where it is 1.5–7%, depending on the sample.

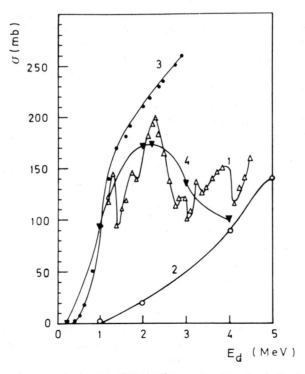

Fig. 2.7 — Excitation function of the $^{12}C(d, n)^{13}N$ reaction, experimentally determined by (1) Jaszczak *et al.* [38], (2) Wilkinson [39], (3) Wohlleben and Schuster [40], and (4) Vandecasteele [41]. For curve 4, σ is given in arbitrary units.

Table 2.6 — $F(E_I)$ for the $^{12}C(d, n)^{13}N$ reaction at 3 MeV. The samples are Al, Ti, Zr and Pb, the standard is graphite. The deviation (%) relative to excitation function 1 is given in brackets.

Excitation function*	$F(E_I)$			
	Al	Ti	Zr	Pb
1	0.757	0.620	0.463	0.278
2	0.769 (+ 1.5)	0.634 (+ 2.1)	0.482 (+ 4.0)	0.293 (+ 5.7)
3	0.758 (+ 0.1)	0.622 (+ 0.3)	0.465 (+ 0.4)	0.279 (+ 0.6)
4	0.757 (− 0.0)	0.621 (+ 0.1)	0.464 (+ 0.3)	0.279 (+ 0.3)

* See Fig. 2.7.

2.3.3.2 Reaction $^{14}N(p, \alpha)^{11}C$

Figures 2.8–2.10 give excitation functions for the $^{14}N(p, \alpha)^{11}C$ reaction, experimentally determined by several authors. With the exception of the excitation function of Epherre and Seide [43] they are in fair agreement.

Table 2.7 gives $F(E_I)$ for 13 MeV protons. The deviation is at most 0.2% and increases with the atomic number of the sample.

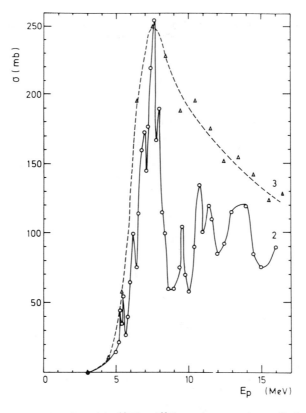

Fig. 2.8 — Excitation function of the $^{14}N(p, \alpha)^{11}C$ reaction, experimentally determined by (2) Jacobs et al. [42], and (3) Epherre and Seide [43].

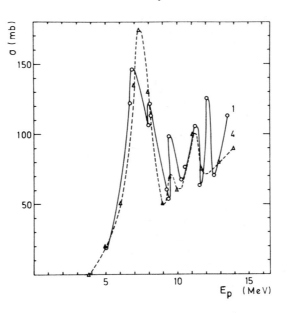

Fig. 2.9 — Excitation function of the ^{14}N(p, α)^{11}C reaction, experimentally determined by (1) Bida *et al.* [44], and (4) Nozaki *et al.* [45].

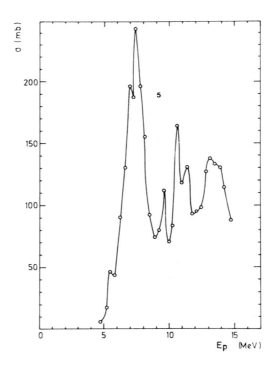

Fig. 2.10 — Excitation function of the ^{14}N(p, α)^{11}C reaction, experimentally determined (5) by Casella *et al.* [46].

Table 2.7 — $F(E_I)$ for the $^{14}N(p, \alpha)^{11}C$ reaction at 13 MeV. The samples are Al, Ti, Zr and Pb, the standard is nylon. The deviation (%) relative to excitation function 2 is given in brackets.

Excitation function*	$F(E_I)$			
	Al	Ti	Zr	Pb
1	0.726 (+ 0.0)	0.638 (+ 0.0)	0.527 (+ 0.1)	0.376 (+ 0.1)
2	0.726	0.638	0.526	0.376
3	0.727 (+ 0.1)	0.638 (+ 0.1)	0.527 (+ 0.1)	0.378 (+ 0.2)
4	0.726 (− 0.0)	0.638 (− 0.0)	0.526 (− 0.0)	0.375 (− 0.1)
5	0.727 (+ 0.0)	0.638 (+ 0.1)	0.527 (+ 0.1)	0.376 (+ 0.1)

* See Figs. 2.8–2.10.

2.3.3.3 Reaction $^{16}O(^3He, p)^{18}F$

Figure 2.11 gives the excitation function of the $^{16}O(^3He, p)^{18}F$ reaction, experimentally determined by Vandecasteele *et al.* [47], Hahn and Ricci [48], and Markowitz and Mahony [49]. The position of the maximum of the curves differs somewhat. Table 2.8 gives $F(E_I)$. The deviations are always smaller than 1%.

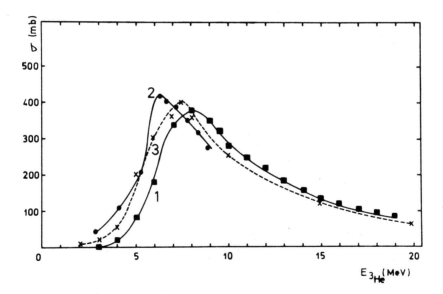

Fig. 2.11 — Excitation function of the $^{16}O(^3He, p)^{18}F$ reaction, determined experimentally by (1) Vandecasteele *et al.* [47], (2) Hahn and Ricci [48], and (3) Markowitz and Mahony [49]. For curve 1, σ is given in arbitrary units.

2.3.3.4 Reaction $^{16}O(\alpha, pn)^{18}F$

Figure 2.12 gives the excitation function of the $^{16}O(\alpha, pn)^{18}F$ reaction, experimentally determined by Vandecasteele *et al.* [47], Furukawa and Tanaka [50], and Rook and Schweikert [51]. The three curves are in fair agreement.

Table 2.9 gives $F(E_I)$. The deviations are all less than 0.5%.

Table 2.8 — $F(E_I)$ for the $^{16}O(^3He, p)^{18}F$ reaction at 9 and 15 MeV. The samples are Al, Ti, Zr and Pb, the standard is quartz. The deviation (%) relative to excitation function 1 is given in brackets.

E_I, MeV	Exci- tation func- tion*	$F(E_I)$			
		Al	Ti	Zr	Pb
9	1	0.898	0.760	0.589	0.379
	2	0.899 (+0.1)	0.760 (0.0)	0.585 (−0.8)	0.377 (−0.6)
	3	0.900 (+0.2)	0.761 (+0.2)	0.585 (−0.6)	0.379 (−0.2)
15	1	0.901	0.768	0.605	0.378
	3	0.901 (+0.04)	0.767 (−0.1)	0.601 (−0.7)	0.375 (−0.9)

* See Fig. 2.11.

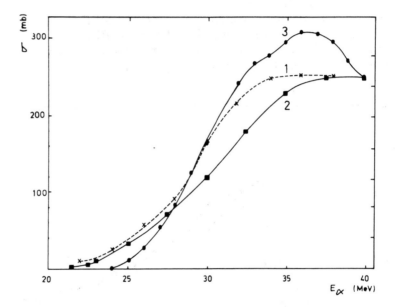

Fig. 2.12 — Excitation function of the $^{16}O(\alpha, pn)^{18}F$ reaction, experimentally determined by (1) Vandecasteele *et al.* [47], (2) Furukawa and Tanaka [50], and (3) Rook and Schweikert [51]. For curves 1 and 3 the cross-section is in arbitrary units.

Table 2.9 — $F(E_I)$ for the $^{16}O(\alpha, pn)^{18}F$ reaction at 30 MeV. The samples are Al, Ti, Zr and Pb, the standard is quartz. The deviation (%) relative to excitation function 1 is given in brackets.

E_I, MeV	Exci- tation func- tion*	$F(E_I)$			
		Al	Ti	Zr	Pb
30	1	0.910	0.792	0.648	0.454
	2	0.910 (+0.0)	0.792 (+0.0)	0.648 (+0.0)	0.454 (+0.1)
	3	0.910 (+0.0)	0.793 (+0.1)	0.649 (+0.2)	0.455 (+0.4)

* See Fig. 2.12.

2.3.3.5 Conclusion

The exact shape of the excitation function has in general only a small influence on $F(E_I)$: for the reactions considered the systematic error due to uncertainties in the excitation function is less than 1%. The systematic error increases with the atomic number of the sample, at least when a standard with a low atomic number is used.

2.3.4 Approximate methods

Table 2.10 gives some frequently used approximate methods for calculating $F(E_I)$ in cases where the excitation function of the nuclear reaction of interest is not known.

Table 2.10 — Approximate standardization methods.

Method			Reference
Ricci and Hahn	$F(E_I) = \dfrac{\Delta R_S}{\Delta R_X}$; $\Delta R = R(E_I) - R(E_T)$	(2.54)	[52]
Ricci and Hahn	$F(E_I) = \dfrac{R_S(E_I)}{R_X(E_I)}$	(2.55)	[53]
Chaudri	$F(E_I) = \dfrac{S_X(E_M)}{S_S(E_M)}$; $E_M = 1/2(E_I + E_T)$	(2.56)	[54]

Figure 2.13(a–d) gives $K(E)$ defined by Eq. (2.57) for aluminium, vanadium, zirconium and lead as samples and for the indicated standard, as a function of the energy.

$$K(E) = \frac{S_X(E)}{S_S(E)} \qquad (2.57)$$

Figure 2.13(a) is for protons, 2.13(b) for deuterons, Fig. 2.13(c) for helium-3 and 2.13(a) for helium-4. The stopping power data of Andersen and Ziegler [19] and of Ziegler [25] were used to calculate $K(E)$ and the curves were normalized, so that $K(E)$ at the maximum energy (25, 15, 34 and 34 MeV, respectively) equals 1. For samples with a low atomic number, for instance aluminium, the curves are practically horizontal over the entire energy range considered. For increasing atomic numbers, $K(E)$ varies more and more with the energy, at least when a material with a low atomic number (graphite, quartz) is used as a standard. For all samples the variation of $K(E)$ is most important at low energy (<7 MeV).

The assumption that, for $E_T < E < E_I$,

$$K(E) = K \qquad (2.58)$$

where K is a constant, is thus a more or less good approximation, depending on E_T, E_I and the atomic numbers of the samples and the standard.

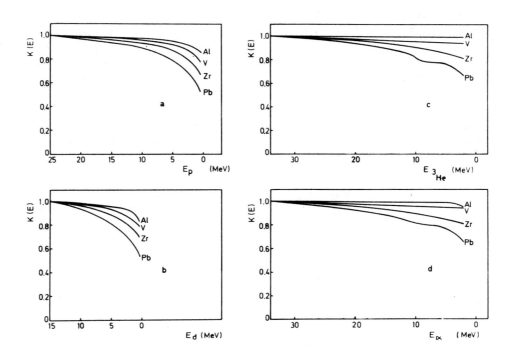

Fig. 2.13 — (a) $K(E)$ as a function of the proton energy. The standard is graphite and the sample Al, V, Zr or Pb. For $E = 25$ MeV, $K(E)$ was set equal to 1.

(b) $K(E)$ as a function of the deuteron energy. The standard is graphite and the sample Al, V, Zr or Pb. For $E = 15$ MeV, $K(E)$ was set equal to 1.

(c) $K(E)$ as a function of the helium-3 energy. The standard is quartz and the sample Al, V, Zr or Pb.

(d) $K(E)$ as a function of the helium-4 energy. The standard is quartz and the sample Al, V, Zr or Pb. For $E = 34$ MeV $K(E)$ was set equal to 1.

It follows from Eqs. (2.53), (2.57) and (2.58) that

$$F(E_I) = K \qquad (2.59)$$

The method of Chaudri *et al.* (Table 2.10) takes for K the value of $K(E)$ at the energy $E_M = \frac{1}{2}(E_I + E_T)$. An upper limit for the systematic error introduced by this approximation can easily be calculated. Indeed, for $E_T < E < E_I$,

$$K(E_T) < K(E) < K(E_I) \qquad (2.60)$$

as appears from Fig. 2.13(a–d). In combination with Eq. (2.53) this yields

$$K(E_T) < F(E_I) < K(E_I) \qquad (2.61)$$

When $F(E_I)$ is approximated by $K(E_M)$, the relative error made in the absolute value is at most

$$\frac{K(E_M) - K(E_T)}{K(E_M)}$$

since

$$K(E_M) - K(E_T) > K(E_I) - K(E_M) \qquad (2.62)$$

as appears from Fig. 2.13(a–d). Table 2.11 gives the maximum systematic error for some typical reactions and for a number of sample–standard combinations. For exoergic reactions $K(E_T)$ was calculated at the Coulomb threshold energy (E_C). For a given reaction, the maximum systematic error increases in general with the atomic number of the matrix. This can be expected since Eq. (2.58) is a better approximation for samples with a low atomic number than for samples with a high atomic number. It appears moreover that for reactions with a high threshold energy the maximum error is in general smaller than for reactions with a low threshold energy, as can also be expected from the shape of the curves giving $K(E)$ as a function of the energy, Fig. 2.13(a–d).

Table 2.11 — Calculated maximum systematic error for Chaudri's method.

Reaction	E_T, MeV	E_I, MeV	Standard	Maximum error, %			
				Al	V	Zr	Pb
$^{12}C(d,n)^{13}N$	0.3	3	Graphite	17.4	7.9	9.0	26.0
$^{14}N(p,n)^{14}O$	6.3	10	Nylon	1.0	1.6	2.7	4.3
$^{14}N(p,\alpha)^{11}C$	3.1	9	Nylon	3.3	5.0	7.9	12.6
$^{14}N(d,n)^{15}O$	$Q>0; E_C=2.2$	5	Nylon	3.6	5.3	7.8	11.5
$^{16}O(^3He,p)^{18}F$	$Q>0; E_C=4.9$	13	Quartz	1.1	1.1	5.6	3.9
$^{16}O(\alpha,pn)^{18}F$	20.4	34	Quartz	0.3	1.1	2.3	4.3

Starting from Eq. (2.58), the validity of the method of Ricci and Hahn [52] can be shown. Since

$$\Delta R = R(E_I) - R(E_T) \int_{E_T}^{E_I} \frac{dE}{S(E)} \qquad (2.63)$$

it follows from

$$K(E) = K \tag{2.64}$$

that

$$\frac{\Delta R_S}{\Delta R_X} = \frac{\displaystyle\int_{E_T}^{E_I} \frac{dE}{S_S(E)}}{\displaystyle\int_{E_T}^{E_I} \frac{dE}{S_X(E)}} = \frac{\displaystyle\int_{E_T}^{E_I} \frac{dE}{S_S(E)}}{\displaystyle\int_{E_T}^{E_I} \frac{dE}{KS_S(E)}} = K \tag{2.65}$$

From Eqs. (2.59) and (2.65), it follows that

$$F(E_I) = \frac{\Delta R_S}{\Delta R_X} \tag{2.66}$$

If it is assumed that $K(E)$ is constant for $0 < E < E_I$, it can be shown in a similar way that:

$$F(E_I) = \frac{R_S(E_I)}{R_X(E_I)} \tag{2.67}$$

This equation was also proposed by Ricci and Hahn [53]. Since it is clear from Fig. 2.13(a–d) that the assumption that $K(E)$ is constant for $0 < E < E_I$, is less justified than for $E_T < E < E_I$, Eq. (2.67) can be expected to be less accurate than Eq. (2.66).

2.3.5 Comparison of approximate methods with numerical integration
In order to compare the standardization methods that do not require knowledge of the (relative) excitation function (Section 2.3.4) with the more exact method using numerical integration (Section 2.3.1), $F(E_T)$ was calculated by these methods for the reactions given in Table 2.12, which gives also the threshold energy, the standard used, and the excitation function used to calculate $F(E_I)$ by numerical integration. The results are given in Figs. 2.14–2.17: the difference (%) between $F(E_I)$ obtained by numerical integration and by the approximate methods is given as a function of the atomic number of the sample for several E_I and for the standard given. The method proposed by Ricci and Hahn [52] gives almost the same result as the one proposed by Chaudri *et al.* [54] and is therefore not shown.

Table 2.12 — Nuclear reactions.

Reaction	E_T, MeV	Standard	Excitation function	
$^{12}C(d, n)^{13}N$	0.3	Graphite	Jaszczak *et al.*	[38]
$^{14}N(p, \alpha)^{11}C$	3.1	Nylon	Jacobs *et al.*	[42]
$^{16}O(^{3}He, p)^{18}F$	$Q > 0; E_C = 4.9$	Quartz	Vandecasteele *et al.*	[47]
$^{16}O(\alpha, pn)^{18}F$	20.4	Quartz	Vandecasteele *et al.*	[47]

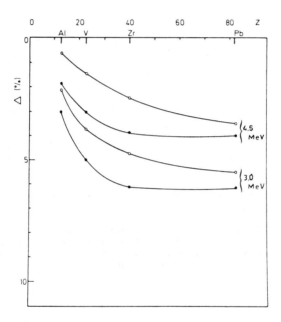

Fig. 2.14 — Per cent difference Δ between $F(E_I)$ obtained by numerical integration, on the one hand, and by the method of Ricci and Hahn [53] (●) or Chaudri *et al.* [54] (○), on the other, as a function of the atomic number Z of the sample. Reaction: $^{12}C(d, n)^{13}N$; standard: graphite. E_I is shown with the curves.

From Figs. 2.14–2.17 the following conclusions can be drawn.

(1) The method proposed by Chaudri *et al.* [54] and the one proposed by Ricci and Hahn [52] are more accurate than the method proposed later on by Ricci and Hahn [53], especially for reactions with a high threshold energy, as already explained in Section 2.3.4.
(2) The systematic error made with the method of Chaudri *et al.* [54] is significantly smaller than the calculated maximum error given in Table 2.11.
(3) The deviation increases with the atomic number of the sample, at least when the standard is an element with a low atomic number or a compound of elements with a low atomic number, lower than that of the sample.

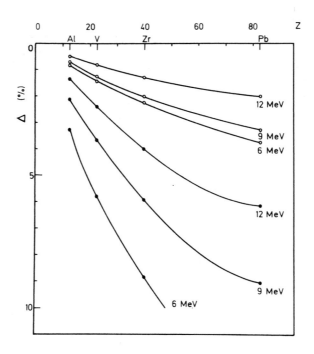

Fig. 2.15 — Per cent difference Δ between $F(E_1)$ obtained by numerical integration, on the one hand, and by the method of Ricci and Hahn [53] (●) or Chaudri et al. [54] (○), on the other, as a function of the atomic number Z of the sample. Reaction $^{14}N(p, \alpha)^{11}C$; standard: nylon. E_1 is shown with the curves.

(4) $F(E_1)$ obtained with the approximate methods is in general smaller than $F(E_1)$ obtained by numerical integration. The approximate methods thus yield a negative error in the cases considered.

2.3.6   Average stopping power method

Ishii et al. [55] developed the average stopping power method. The 'average stopping power' $\langle S \rangle$ is defined by Eq. (2.68)

$$\int_0^{E_1} \frac{\sigma(E)dE}{S(E)} = \frac{1}{\langle S \rangle} \int_0^{E_1} \sigma(E)dE \qquad (2.68)$$

An 'average energy' E_m is defined by Eq. (2.69):

$$\langle S \rangle = S(E_m) \qquad (2.69)$$

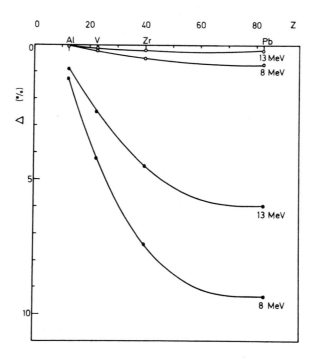

Fig. 2.16 — Per cent difference Δ between $F(E_1)$ obtained by numerical integration, on the one hand, and by the method of Ricci and Hahn [53] (●) or Chaudri *et al.* [54] (○), on the other, as a function of the atomic number Z of the sample. Reaction: $^{16}O(^3He, p)^{18}F$; standard = quartz. E_1 is shown with the curve.

so

$$\frac{1}{S(E_m)} = \frac{\int_0^{E_1} \dfrac{\sigma(E)dE}{S(E)}}{\int_0^{E_1} \sigma(E)dE} \qquad (2.70)$$

By using the Bethe equation (2.16) in the non-relativistic form and without taking into account the shell-correction term, it can be shown that E_m is approximately given by

$$E_m = \frac{\int_0^{E_1} E\sigma(E)dE}{\int_0^{E_1} \sigma(E)dE} \qquad (2.71)$$

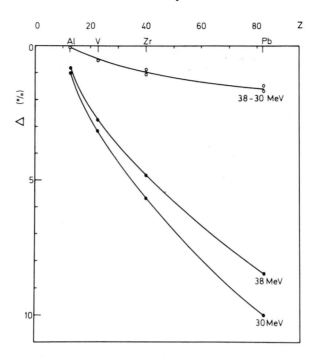

Fig. 2.17 — Per cent difference Δ between $F(E_I)$ obtained by numerical integration, on the one hand, and by the method of Ricci and Hahn [53] ($\bullet$) or Chaudri *et al.* [54] ($\bigcirc$), on the other, as a function of the atomic number Z of the sample. Reaction: $^{16}O(\alpha, pn)^{18}F$; standard = quartz. E_I is shown with the curves.

E_m is thus, to a good approximation, independent of the matrix and depends only on the nuclear reaction and on the incident energy E_I. $F(E_I)$ is given by

$$F(E_I) = \frac{S_X(E_m)}{S_S(E_m)} \tag{2.72}$$

Ishii *et al.* [55] also give equations that allow an estimate of the error in $F(E_I)$ thus obtained.

In practice, for a given nuclear reaction and incident energy E_I, E_m is calculated by means of Eq. (2.71) and $F(E_I)$ by means of Eq. (2.72). This method has the practical advantage that E_m can be calculated once and for all, so that calculating $F(E_I)$ by Eq. (2.72) for a given standard–sample pair is rapid. For its application, this method requires, just as numerical integration does, knowledge of the excitation function. The average stopping power method introduces, however, additional approximations.

In a later paper, Ishii *et al.* [56] described a method for calculating E_m from an activation curve, i.e. the curve that gives the thick-target yield for a given nuclear reaction as a function of the incident energy.

2.3.7 Conclusion

When the excitation function of the reaction of interest is known, the numerical integration method or the average stopping power method are the preferred methods for calculating $F(E_I)$. When this is not the case, the approximate methods proposed by Ricci and Hahn [52] or by Chaudri et al. [54] can be used. These methods yield a negligible systematic error for reactions with a high threshold energy and for samples with a low atomic number, if a standard with a low atomic number is used.

2.4 NUCLEAR INTERFERENCES

The determination of element A in matrix X by using the nuclear reaction A(a, b)C (threshold energy E_T) may be interfered with by the nuclear reaction B(a, b')C (threshold energy E'_T), both reactions yielding the same radionuclide C. The interference is called nuclear interference.

Nuclear interference is often avoided by choosing an incident energy E_I so that $E_T < E_I < E'_T$. If $E_T > E'_T$ this is of course impossible, but, when the concentration of the interfering element B is known, the interference can be corrected for.

When the matrix X is irradiated, the counting rate of C at the end of the irradiation is given by Eq. (2.73), which follows from Eqs. (2.46) and (2.47).

$$a_X = i_X(1 - e^{-\lambda t_b, x})(k_A c_{A,X} INT_{A,X} + k_B c_{B,X} INT_{B,X}) \qquad (2.73)$$

where a = count-rate at the end of the irradiation, i = beam intensity, λ = disintegration constant of C, t_b = irradiation time, k = proportionality constant, c = concentration and

$$INT = \int_0^{E_I} \frac{\sigma(E) dE}{S(E)} \qquad (2.74)$$

A and B refer to the elements A and B or to the nuclear reactions A(a, b)C and B(a, b')C, and X to the matrix.

Equation (2.73) can be written as:

$$a_X = k_A i_X(1 - e^{-\lambda t_b, x}) INT_{A,X}(c_{A,X} + IF c_{B,X}) \qquad (2.75)$$

or

$$a_X = k_A i_X(1 - e^{-\lambda t_b, x}) INT_{A,X} c'_{A,X} \qquad (2.76)$$

where

$$IF = \frac{k_B}{k_A} \frac{INT_{B,X}}{INT_{A,X}} \tag{2.77}$$

and

$$c'_{A,X} = c_{A,X} + IFc_{B,X} \tag{2.78}$$

IF is the interference factor of element B with the determination of element A in the matrix X. $c'_{A,X}$ is the apparent concentration of A, i.e. the concentration obtained when the interference of B is not taken into account.

IF is determined as follows: a standard (SA) containing element A and not B and a standard (SB) containing B and not A are both irradiated at an incident energy E_I and measured under the same conditions. The count-rate at the end of the irradiation is given by:

$$a_{SA} = k_A i_{SA}(1 - e^{-\lambda_{tb,SA}})c_{A,SA}INT_{A,SA} \tag{2.79}$$

$$a_{SB} = k_B i_{SB}(1 - e^{-\lambda_{tb,SB}})c_{B,SB}INT_{B,SB} \tag{2.80}$$

IF can then be calculated by means of:

$$IF = \frac{a_{SB}}{a_{SA}} \frac{i_{SA}}{i_{SB}} \frac{(1 - e^{-\lambda_{tb,SA}})}{(1 - e^{-\lambda_{tb,SB}})} \frac{c_{A,SA}}{c_{B,SB}} \frac{INT_{A,SA}}{INT_{B,SB}} \frac{INT_{B,X}}{INT_{A,X}} \tag{2.81}$$

Under the same conditions as in Section 2.3.4, use can be made of approximate methods to calculate IF. IF is for instance approximately given by:

$$IF = \frac{a_{SB}}{a_{SA}} \frac{i_{SA}}{i_{SB}} \frac{(1 - e^{-\lambda_{tb,SA}})}{(1 - e^{-\lambda_{tb,SB}})} \frac{c_{A,SA}}{c_{B,SB}} \frac{R_{SA}(E_I)}{R_{SB}(E_I)} \tag{2.82}$$

REFERENCES

[1] D. De Soete, R. Gijbels and J. Hoste, *Neutron Activation Analysis*, Wiley–Interscience, London, 1972.
[2] P. J. Elving, V. Krivan and I. M. Kolthoff (eds.), *Treatise on Analytical Chemistry*, Part I, 2nd Ed., Vol. 14, Wiley–Interscience, New York, 1986.
[3] J. Tölgyessy and I. M. Kyrš. *Nuclear Analytical Chemistry*, Horwood, Chichester, in the press.
[4] N. Seelmann-Eggebert, G. Pfennig, H. Muenzel and H. Klewe-Nebenius, *Karlsruher Nuklidkarte*, 5th Ed., Kernforschungzentrum, Karlsruhe, 1981.
[5] K. A. Keller, J. Lange and H. Muenzel, Q-*Values*, in *Landolt-Börnstein Numerical Data and Functional Relationships in Science and Technology*, K. Hellwege and H. Schopper (eds.), Group I, Vol. 5, Part a, Springer-Verlag, Berlin, 1973.
[6] M. Lefort, *Nuclear Chemistry*, Van Nostrand, London, 1968.
[7] G. Friedlander, J. W. Kennedy, E. S. Macias and J. M. Miller, *Nuclear and Radiochemistry*, Wiley, New York, 1981.
[8] G. M. Temmer, *Phys. Rev.*, 1949, **76**, 424.
[9] J. W. Meadows, *Phys. Rev.*, 1953, **91**, 885.
[10] K. A. Keller, J. Lange, H. Münzel and G. Pfennig, *Excitation Functions for Charged Particle Induced Nucelar Reactions*, in *Landolt-Börnstein Numerical Data and Functional Relationships in Science and Technology*, K. Hellwege and H. Schopper (eds.), Group I, Vol. 5, Part b, Springer-Verlag, Berlin, 1973.
[11] F. K. McGowan and W. T. Milner, *Charged Particle Reaction List* 1948–1971, K. Way (ed.), Academic Press, New York, 1973.
[12] U. Fano, *Annual Review of Nuclear Science*, 1963, **13**, 1.
[13] H. H. Andersen, H. Sorensen and P. Vadja, *Phys. Rev.*, 1969, **180**, 373.
[14] W. Brandt, *Phys. Rev.*, 1958, **112**, 1624.
[15] E. Bonderup, *Kgl. Danske Videnskab. Selskab., Mat.-Fys. Medd.*, 1967, **35**, No. 17.
[16] M. Walske, *Phys. Rev.*, 1952, **88**, 1283.
[17] M. Walske, *Phys. Rev.*, 1957, **101**, 940.
[18] J. Lindhard and A. Winther, *Kgl. Danske Videnskab. Selskab., Mat.-Fys. Medd.*, 1964, **34**, No. 4.
[19] H. H. Andersen and J. F. Ziegler, *Hydrogen, Stopping Powers and Ranges in all Elements*, Pergamon Press, New York, 1977.
[20] P. Vavilov, *Soviet Physics*, 1957, **5**, 749.
[21] G. Deconninck, *Introduction to Radioanalytical Physics*, Elsevier, Amsterdam, 1978.
[22] N. Bohr, *Phil. Mag.*, 1915, **30**, 581.
[23] M. D. Tran and J. Tousset, in *Modern Trends in Activation Analysis*, J. R. De Voe (ed.), Vol. II, p. 754. NBS, Washington DC, 1969.
[24] W. Burcham, *Nuclear Physics*, Longmans, London, 1970.
[25] J. F. Ziegler, *Helium, Stopping Powers and Ranges in all Elemental Matter*, Pergamon, New York, 1977.
[26] J. Janni, *Report* AFWL-TR65-150, 1966.
[27] H. Andersen, *Studies of Atomic Collisions in Solids by Means of Calorimetric Techniques*, University of Aarhus, Aarhus, Denmark, 1974
[28] W. Chu and D. Powers, *Phys. Lett.*, 1972, **A40**, 23.
[29] H. Bichsel, *American Institute of Physics Handbook*, 2nd Ed., McGraw-Hill, New York, 1963.
[30] J. Lindhard and M. Scharff, *Phys. Rev.*, 1961, **124**, 128.
[31] C. Varelas and J. Biersack, *Nucl. Instr. Meth.*, 1970, **79**, 213.
[32] W. Bragg and R. Kleeman, *Phil. Mag.*, 1905, **10**, 318.
[33] R. Langley and R. Blewer, *Nucl. Instr. Meth.*, 1976, **132**, 109.
[34] K. Ishii, G. Blondiaux, M. Valladon and J. L. Debrun, *Nucl. Instr. Meth.*, 1979, **158**, 199.
[35] G. Blondiaux, M. Valladon, K. Ishii and J. L. Debrun, *Trans. Am. Nucl. Soc.*, 1979, **32**, 199.
[36] G. Blondiaux, M. Valladon, K. Ishii and J. L. Debrun, *Nucl. Instr. Meth.*, 1980, **168**, 29.
[37] J. Baglin and J. Ziegler, *J. Appl. Phys.*, 1974, **45**, 1413.
[38] R. J. Jaszczak, R. L. Macklin and J. H. Gibbons, *Phys. Rev.*, 1969, **181**, 1428.
[39] D. H. Wilkinson, *Phys. Rev.*, 1955, **100**, 32.
[40] K. Wohllebran and E. Schuster, *Radiochim. Acta*, 1967, **8**, 78.
[41] C. Vandecasteele, Unpublished results.
[42] W. W. Jacobs, D. Bodansky, D. Chamberlin and D. L. Oberg, *Phys. Rev. C*, 1974, **9**, 2134.
[43] M. Epherre and C. Seide, *Phys. Rev. C*, 1971, **3**, 2167.
[44] G. T. Bida, T. J. Ruth and A. P. Wolf, *Radiochim. Acta*, 1980, **27**, 181.
[45] T. Nozaki, M. Okuo, H. Akutzu and M. Furukawa, *Bull. Chem. Soc. Japan*, 1966, **39**, 2685.
[46] V. R. Casella, D. R. Christman, T. Ido and A. P. Wolf, *Radiochim. Acta*, 1978, **25**, 17.

[47] C. Vandecasteele, F. Adams and J. Hoste, *Anal. Chim. Acta*, 1974, **71**, 67.
[48] R. L. Hahn and E. Ricci, *Phys. Rev.*, 1966, **146**, 650.
[49] S. S. Markowitz and J. D. Mahony, *Anal. Chem.*, 1962, **34**, 329.
[50] M. Furukawa and S. Tanaka, *J. Phys. Soc. Japan*, 1961, **16**, 129.
[51] H. L. Rook and E. A. Schweikert, *Anal. Chem.*, 1969, **41**, 958.
[52] E. Ricci and R. L. Hahn, *Anal. Chem.*, 1965, **37**, 742.
[53] E. Ricci and R. L. Hahn, *Anal. Chem.*, 1967, **39**, 794.
[54] M. A. Chaudri, G. Burns, E. Reen, J. L. Rouse and B. M. Spicer, *Proc. Int. Conf. Mod. Trends in Activation Analysis*, Munich, 1976.
[55] K. Ishii, M. Valladon and J. L. Debrun, *Nucl. Instr. Meth.*, 1978, **150**, 213.
[56] K. Ishii, M. Valladon, C. S. Sastri and J. L. Debrun, *Nucl. Instr. Meth.*, 1978, **158**, 503.

3

Apparatus and experimental technique

3.1 IRRADIATION

3.1.1 Accelerator

The acceleration of charged particles to energies sufficiently high to induce nuclear reactions was first achieved by means of a high voltage across an accelerating tube. In order to obtain an accelerating voltage above 200 kV, a voltage multiplier rectifier as first developed by Cockroft and Walton may be used. Voltages up to a few MV and currents up to 10 mA for protons can be obtained. Because of the rather limited energy, these generators are rarely used in charged-particle activation analysis. They are, however, very suitable as 14 MeV neutron generators. In a 14 MeV neutron generator deuterons accelerated to a few 100 keV bombard a tritium target, i.e. tritium bound to titanium or zirconium. The t(d, n)α reaction, with a Q-value of 17.6 MeV, takes place and yields neutrons of 14.9, 14.1 and 13.3 MeV, respectively, at an angle of 0, 90 and 180° relative to a 150 keV deuteron beam. The fact that this reaction has a high cross-section at low deuteron energy (maximum of the excitation function at 110 keV) is, of course, a consequence of the low Coulomb barrier energy.

Of much more use in CPAA is the Van de Graaff electrostatic generator, developed by Van de Graaff from 1929 onwards. This belt-driven electrostatic generator allows a constant potential V of up to about 15 MV to be obtained, and thus allows acceleration of particles, of charge q, to an energy qV.

Because the potentials applicable in electrostatic generators are limited by insulation problems, multistage acceleration must be used to obtain still higher energies. The only accelerator applying multistage acceleration that is of practical significance in CPAA is the cyclotron, proposed by Lawrence in 1929. Whereas in classical cyclotrons, the energy is limited by the relativistic increase of the mass of the accelerated particle, this limitation is overcome in isochronous or sector-focused cyclotrons.

3.1.1.1 Van de Graaff electrostatic generator

In the Van de Graaff machine a high potential is built up and maintained on a conducting terminal by the continuous transfer of static charge from a moving charging system to the terminal. It consists of the following items.

(1) A terminal at high potential.
(2) A pressure vessel in which the terminal and its supporting column are enclosed. The pressure vessel is filled with an insulating gas, e.g. 80% N_2 and 20% CO_2, pure SF_6, or $N_2 + CO_2$ with small additions of SF_6. A typical gas pressure is 7 bar.
(3) A column supporting the high voltage terminal and providing a region of uniform potential gradient within which the accelerating tube and charging system are contained. The column consists of a large number of equipotential plates, separated by insulating spacers and connected by high voltage resistors.
(4) A charging system. The most common charging system is still the charging belt, originally used by Van de Graaff. These belts usually consist of a woven cotton base coated with synthetic rubbber and may be charged directly or by induction. A new type of charging system is the so-called pelletron, the charging chain of which consists of steel conductors separated by nylon spacers. Its advantages are reproducible mechanical properties, long-life and very stable operating conditions.
(5) An accelerating tube, constructed from metal electrodes spaced by ceramic or glass insulators. Each electrode is connected to the corresponding equipotential plate, and the electrodes are shaped so that they shield the insulators.
(6) An ion source, usually of the radiofrequency type, situated in the high-voltage terminal.
(7) A voltage stabilization system. The terminal voltage can readily be changed by changing the loading current. The reference for the stabilization system can be the signal taken from the jaws of the exit slit situated behind a 90° beam-analysing magnet following the accelerator, or the signal from a generating voltmeter.

Typical characteristics for a single-ended generator are: terminal voltage: 2–10 MV; current: 0.1–0.5 mA. Voltages up to 15 MV have been obtained, however.

The major advantage of electrostatic generators is that they supply ions of precisely controllable energy: constant to about 0.1% and with an energy spread of about the same size.

In a tandem generator the column structure runs right through the pressure tank and is divided symmetrically by the high-voltage terminal. Negative ions of low energy (~ 100 keV) are injected from an external source and accelerated toward the terminal. The negative ion beam then passes through a thin foil or through a tube filled with gas at low pressure, where the ions are stripped of part of their electrons, so that positive ions emerge and are further accelerated to ground potential. Protons thus acquire an energy corresponding to twice the terminal voltage. For heavier ions this may be $z + 1$ times the terminal voltage, where z is the positive charge during the second stage of acceleration, expressed as the number of elementary charges.

Both horizontal and vertical tandems have been constructed. Two-stage tandems producing proton beams with an intensity of 10–50 μA and an energy of about 25 MeV are commercially available. Tandems with a terminal voltage of up to 20 MV are under construction.

More detailed discussions of electrostatic generators are given by Livingston and Blewet [1] and Allen [2].

3.1.1.2 Cyclotron

In a cyclotron charged particles are accelerated to an energy of betwen a few MeV and a few hundred MeV. The particles are constrained by a magnetic field to move in circular orbits, so that they repeatedly traverse the same electrical potential difference, for instance 10 kV. A cyclotron consists essentially of two hollow metal accelerating electrodes, between which an oscillating electric field is generated by a radiofrequency oscillator. The electrodes, called dees from their shape, are placed in a uniform magnetic field (Fig. 3.1). In an ion source, located centrally between the

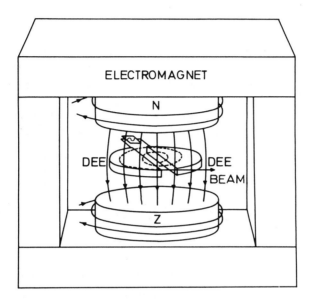

Fig. 3.1 — Classical cyclotron.

dees, charged particles are attracted by the negatively charged dee and describe, under the influence of the magnetic field, a semi-circular trajectory inside the dee. When they again reach the gap between the dees, the potential is reversed, leading to a second acceleration. Because of the higher velocity a semicircle with a larger radius is described. The particles are accelerated at each gap crossing, and the radius of the orbit increases with the velocity, so that the particles follow a spiral path from the ion source to the edge of the magnet, where they are pulled out as an external beam by the action of an electrostatic deflector.

It is fundamental to the operation of the cyclotron that the time required for one orbit is independent of the radius of the orbit and of the velocity. Particles that cross the gap between the dees at such a time that the potential difference is capable of accelerating them will continue to be accelerated by all subsequent gap crossings, provided that the charge to mass ratio, q/m, does not change and that the frequency and the magnetic field are constant.

Since the Lorentz force on the particle is equal to the centrifugal force, the angular velocity ω is given by Eq. (3.1).

$$\omega = qB/m \tag{3.1}$$

where q = charge; B = magnetic induction; m = mass.

Between the dees an alternating voltage with a frequency v given by Eq. (3.2) must be applied

$$\omega = 2\pi v \tag{3.2}$$

In a totally uniform magnetic field the orbits of the particles are unstable in the vertical plane, so a particle with a small velocity component in the vertical (axial) direction will soon strike the inside of a dee and be lost. In order that a particle deviating vertically from its path in the median plane will be returned to this plane, the magnetic field must decrease with the radius. The magnetic field has then, outside the median plane, a radial component B_r (Fig. 3.2), directed so that a vertical force is

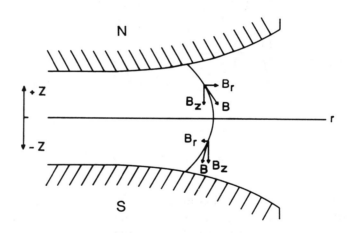

Fig. 3.2 — Vertical focusing in a classical cyclotron.

exerted to bring the particle back to the median plane. Since B_r is approximately proportional to the distance Z from the median plane, (damped) harmonic oscillations around this plane occur. Classical cyclotrons were constructed in this way. When the magnetic field decreases with the radius, it follows from Eq. (3.1) that ω is not constant. The decrease of the magnetic field is, however, kept small, so that acceleration remains possible during a large number of turns.

According to special relativity, the mass m of a particle is related to its rest-mass m_0 by Eq. (3.3):

$$m = m_0 \left[1 - \frac{v^2}{c^2} \right]^{-1/2} \tag{3.3}$$

where v = particle velocity; c = speed of light.

The mass thus increases during acceleration. According to Eq. (3.1), the angular velocity is smaller for a fast particle than for a slow particle. If the accelerating voltage is in phase with a slow particle, it is not in phase with a more energetic particle. The particles thus continuously get out of phase, so crossing the gap between the dees finally leads to deceleration. The classical (constant frequency, slightly falling magnetic field) version of the cyclotron is therefore limited to the acceleration of particles to modest energy, or a kinetic energy of about 12 MeV for protons, for a dee-to-dee voltage of 100 kV).

Although classical cyclotrons were very valuable instruments in their day, few of these machines now remain in service. These machines were rather simple to operate and capable of producing high beam intensities, up to several hundreds of μA.

Two methods have been devised to overcome the relativistic limit.

(1) In the frequency-modulated (FM) cyclotron the magnetic field decreases with the distance from the centre, in order to preserve axial stability. When a group of particles is accelerated, the frequency is reduced so that the particles remain in phase with the accelerating potential. As soon as the acceleration of a group of particles is complete, the frequency is restored to its initial value and a second group of particles is accelerated. The beam of the FM cyclotron consists therefore of pulses appearing at the repetition rate of the frequency modulation (typically 100 pulses/sec), whereas a classical cyclotron produces pulses at the oscillator frequency (typically 10^7 pulses/sec). The average mean intensity is 1–2 orders of magnitude lower than for a classical cyclotron and the largest machines allow acceleration of protons up to nearly 1 GeV.

(2) In the azimuthally varying field (AVF) cyclotron the magnetic field increases with the distance from the centre, in order to compensate for the relativistic increase of the particle mass. This is usually accomplished by means of circular coils placed on the surface of the magnet poles. Axial stability is obtained by adding ridges to the poles, so that a particle in orbit moves in a magnetic field that is alternately higher and lower than the average: regions of high field (hills) alternate with regions of low field (valleys). The focusing action is made even stronger if the sectors are given a spiral shape. Since the time required for an orbit is constant, independent of the particle velocity, the AVF machine is often referred to as an isochronous cyclotron. The name sector-focused cyclotron is also often used.

In recent years no other types of cyclotrons have been constructed. The AVF cyclotron produces beams with similar intensity and time structure to the classical cyclotron, but much higher energies. up to 600 MeV, are reached in the largest machines. Even for relatively modest energies (10–20 MeV for protons) the AVF design has the advantage of allowing much lower dee-to-dee voltages, resulting in a more compact accelerator. When an oscillator with variable frequency is used, particles may be accelerated to any final energy, from very low to a maximal value. In addition to the oscillator frequency, it is necessary to adjust the average magnetic field (by changing the current in the main coils) and the radial shape of the magnetic field (by changing the current in the circular coils on the pole faces) for each particle and each energy, in order to satisfy Eq. (3.1). Most AVF cyclotrons are multi-particle and variable-energy machines.

A typical compact isochronous cyclotron is the 520 type of the CGR-MeV (Buc, France). Table 3.1 gives some characteristics of this accelerator.

Table 3.1 — Characteristics of the CGR-MeV 520 cyclotron.

Electromagnet	
Pole diameter	1.20 m
Number of sectors	4
Average magnetic field at	
extraction radius	1.48 T
Extraction radius	0.525 m
Particle energies p	2.5–24 MeV
d	3–14.5 MeV
³He	6–32 MeV
⁴He	10–29 MeV
Maximum intensity of	
extracted beam	
p	100 μA
d	100 μA
³He	60 μA
⁴He	60 μA

A more detailed treatment of the cyclotron is given by Livingood [3], Livingston and Blewet [1], and Harvey [4].

3.1.1.3   *Choice of an accelerator*
Particles of the energy of interest for CPAA, roughly 2–20 MeV for protons and deuterons and 10–40 MeV for helium-3 and helium-4, can be obtained by means of a (tandem) electrostatic generator or a cyclotron.

For low energy work (<5 MeV for protons, deuterons or tritons, <10 MeV for helium-3 or helium-4) the single-ended electrostatic generator is to be preferred, as it is easy to operate and very reliable. Of course, some compact isochronous cyclotrons can also be used at these energies, but they are more complex machines. On the other hand, (tandem) electrostatic generators with terminal voltages above 5 MV are in general larger and more complex machines than compact cyclotrons. Their major advantage, namely that of yielding particles of precisely controllable energy, is seldom of importance in CPAA (see Section 3.5). At energies above 5 MeV for protons and above 10 MeV for helium ions, (compact) isochronous cyclotrons are therefore the preferred accelerators. The machines are rather easy to operate and very versatile and allow several other applications, for instance:

— production of radionuclides for medical use:
— proton-induced X-ray emission (PIXE);
— fast neutron activation analysis.

In recent years, the number of compact isochronous cyclotrons has considerably increased.

3.1.2 Irradiation set-up and sample holder

After extraction the charged particle beam is sent through a set of evacuated metal tubes (typical vacuum $= 10^{-4}$ Pa) onto the target. The beam transport systems in use vary from very simple to very complex. In simple systems only one target is placed in front of the accelerator, in more complex systems up to some 10 irradiation positions are available, situated in concrete-shielded rooms. The beam transport system usually comprises one or several steering magnets to send the beam in one of several directions, quadrupole magnets to focus the beam, and Faraday cups, i.e. insulated electrodes that allow the beam to be stopped and its intensity measured before it is sent on to the target. A convenient way to check the position and dimensions of the beam is to stop it on a remotely controlled aluminium oxide plate, just in front of the target, and to observe the image by means of a closed TV-circuit.

The irradiation set-up and the target holders vary from one installation to another. Therefore those in use at the cyclotron of Ghent University, Belgium, will be discussed, as a typical example. Figure 3.3 shows the end of the beam line for

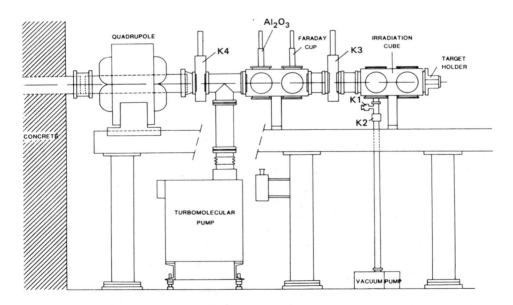

Fig. 3.3 — End of the beam line for activation analysis with the cyclotron of Ghent University.

activation analysis. Before impinging on the sample the beam passes through a collimator. The collimators in use are 8 cm long and have internal diameters of 0.8, 1.2 and 1.7 cm, respectively. Their axis coincides with that of the tube of the beam transport system.

The sample is placed in an electrically insulated target holder, cooled with demineralized water and mounted on the experimental set-up. An O-ring serves as electrical insulation and as a vacuum seal. Figure 3.4 shows a simple sample holder

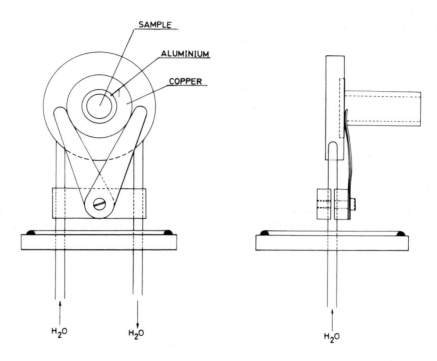

Fig. 3.4 — Sample holder in use with the cyclotron of Ghent University.

suitable for the irradiation of disc-shaped samples (13–20 mm diameter, a few mm thick). An aluminium tube is placed on the target holder in order to reduce errors caused in the beam intensity readings by escape of secondary electrons. Cooling of the irradiated samples is not very effective, so this sample holder is only suitable for use at low beam intensities ($< 1 \mu A$).

Figure 3.5 shows a more complicated sample holder [6], also suitable for the irradiation of disc-shaped samples. The cooling of the sample is more effective, so irradiations at higher intensities can be carried out. A water-cooled collimator is placed on the target holder.

In addition to massive samples, powdered samples are also often analysed by CPAA. Figure 3.6 shows an aluminium sample holder developed [7] for the irradiation of powdered rock samples. A thin aluminium foil and a suitable metal foil, serving as a beam intensity monitor, are placed in C and kept in place by means of tube B which is attached to C with the ring A. The sample is brought into C from the back by means of a spatula. D is inserted into C and kept in place by E. During the irradiation this sample holder is placed in the target holder shown in Fig. 3.4. After

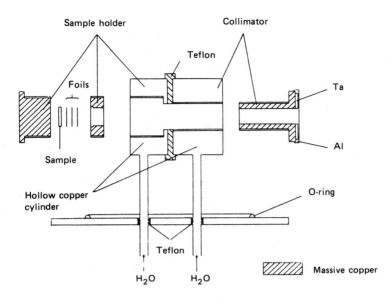

Fig. 3.5 — Sample holder in use with the cyclotron of Ghent University [6].

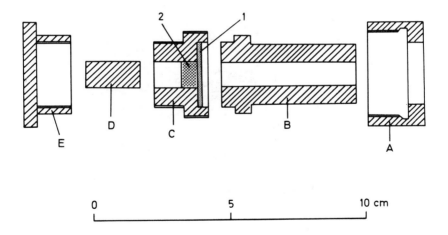

Fig. 3.6—Sample holder for the irradiation of powdered samples: 1, nickel and aluminium foil; 2, sample. For the explanation of symbols A–E, see text.

the irradiation A and B are detached from C and the beam intensity monitor is removed. E is screwed further onto C, to recover the sample quantitatively.

After introduction of the sample holder into the experimental set-up (Fig. 3.3), valve K1 (air inlet) is closed, valve K2 is opened, and a primary mechanical pump is used to pump down the irradiation cube. As soon as the pressure is below 5 Pa, K2 is closed and K3 opened. As soon as the pressure between K3 and K4 is below 10^2Pa, K4 is opened and the irradiation can be started by lifting a Faraday cup from the beam.

When powdered samples with low thermal conductivity, e.g. rocks or environmental materials, are irradiated, systematic errors may occur owing to volatilization of the analyte element or of matrix components. Wauters et al. [8] have shown that for irradiations under helium gas the temperature obtained is considerably lower than that under vacuum. Figure 3.7 shows the target system used, which has the following essential features.

(1) The samples are irradiated under helium instead of under vacuum, resulting in improved heat transport.
(2) The sample holder is efficiently cooled with demineralized water.
(3) Loading, irradiation and quantitative recovery of the sample are easy.

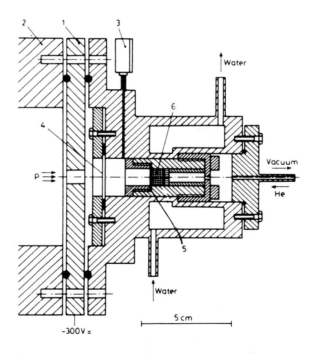

Fig. 3.7 — Target system for the irradiation of powdered samples under helium: 1, diaphragm; 2, aluminium block; 3, manometer; 4, titanium foil; 5, beam intensity monitor foil; 6, sample.

An electrically insulated diaphragm, 8 mm in diameter (1) is placed between the aluminium block (2) terminating the beam transport system, and the sample holder. The beam diameter is larger than the diameter of the diaphragm, so the samples are irradiated with an 8 mm diameter beam. A negative potential (~ 300 V) on the diaphragm prevents secondary electrons from escaping, which would invalidate the current measurement. The aluminium sample-holder is electrically insulated from the diaphragm with a Vitron O-ring. The sample holder consists of two parts: one remains fixed onto the aluminium block; the other, which holds the sample, can easily be removed. The sample holder is filled with helium at a pressure controlled by means of a manometer (3). A 50-μm thick titanium foil (4) separates the helium from the vacuum in the beam transport system. The removable part of the sample holder is shown in Fig. 3.8. A beam intensity monitor (5) is kept in place by means of part A.

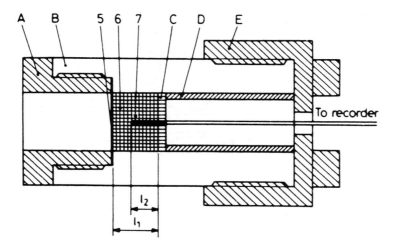

Fig. 3.8 — Removable part of the sample holder: 5, beam intensity monitor foil; 6, sample; 7, thermocouple; for A, B, C, D, E see text.

The powdered sample (6) is introduced into part B, and a sintered polyethylene disc C is placed on top of it and kept in place by means of parts D and E. The helium diffuses through the sintered polyethylene disc into the sample. After the sample has been loaded, the removable part of the sample holder is brought into place, the sample holder is evacuated, and, when necessary, helium is introduced at the required pressure. In Fig. 3.8 the thermocouple used for studying the temperature during the irradiation (see Section 6.1) is also shown.

In some analyses use is made of short-lived radionuclides ($t_{1/2} < 5$ min). The sample must then, of course, be measured as soon as possible after the irradiation. In addition, directly after the irradiation, the radiation level in the irradiation area may be high, especially after deuteron irradiation. From the aluminium parts of the beam transport system and the collimator ^{28}Al (β^-, γ-emitter, $t_{1/2} = 2.24$ min) is produced by the ^{27}Al(d, p)^{28}Al reaction. Entering the irradiation area and removal of the sample directly after the irradiation must thus be avoided. For these reasons use can be made of a pneumatic transfer system, for disc-shaped samples mounted on a copper 'rabbit'. The sending station is situated outside the concrete-shielded irradiation area, close to the measuring room.

The transfer of the sample and production of the vacuum are microprocessor-controlled. Figure 3.9 schematically shows the system. Initially, all electro-pneu-

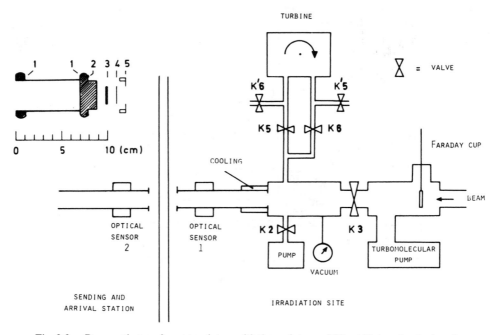

Fig. 3.9 — Pneumatic transfer system in use with the cyclotron of Ghent University: 1, rings for transport and thermal contact with cooling mantle; 2, O-ring; 3, sample; 4, monitor foil; 5, ring with screw thread.

matic valves (K) are closed and the vacuum pump and turbine are off. The following sequence is carried out under microprocessor control:

(1) turbine on and valves K5 and K'5 open;
(2) rabbit passes optical sensors 2 and 1;
(3) K5 and K'5 closed, turbine off;
(4) vacuum pump on and valve K2 open;
(5) as soon as the pressure is below 5 Pa, K2 closed, vacuum pump off, K3 open, Faraday cup deblocked.

The sample is ready to be irradiated as soon as a vacuum of 10^{-2} Pa is reached.

During the irradiation the 'rabbit' is in contact (through two copper rings) with a cooling mantle through which demineralized water flows.

After the preset irradiation time the irradiation is stopped by lowering the Faraday cup. The following sequence is then immediately carried out:

(1) valve K3 closed, turbine on and valves K6 and K'6 open;
(2) rabbit passes optical sensors 1 and 2;
(3) valves K6 and K'6 closed and turbine off.

About 10 sec after the end of the irradiation the sample is available for further treatment.

3.1.3 Determination of the beam intensity

It appears from Eq. (2.49) that the beam intensities for the sample and the standard, or at least their ratio, must be known in order to calculate the concentration of the sample. It is, in general, assumed that the beam intensity is constant during the irradiation.

The beam intensity can be determined directly by measurement of the electric current incident on the sample and the standard. The electric current I (μA) and the beam intensity i (particles/sec) are related by Eq. (3.4):

$$i = 6.241 \times 10^{12} \, I/z \tag{3.4}$$

where z = charge of the incident particle, expressed as the number of elementary charges.

In order to allow measurements of the incident current, the sample holder must be electrically insulated from the experimental set-up, which is grounded. The cooling liquid must also have a sufficiently low specific conductivity: demineralized water is frequently used as the cooling liquid. In front of the sample holder a collimator is placed, in order to eliminate the part of the beam that falls outside the sample. Since emission of secondary electrons may result in inaccurate beam intensity measurements, an insulated ring-shaped electrode at a negative potential, for instance -300 V, is usually placed in front of the sample.

Direct current measurement has the following disavantages.

(1) Emission of secondary electrons can lead to inaccurate measurements, and an insufficient vacuum can, because of ionization, falsify the measurement.
(2) At low intensities an accurate measurement may be difficult.

The beam intensity can also be determined indirectly by using beam intensity monitor foils. During the irradiation a metal foil with a thickness smaller than the range is placed before the sample and the standard. The activity of the monitor foil at the end of the irradiation, A_M, is given by Eq. (3.5)

$$A_M = n_M i(1 - e^{-\lambda_M t_b}) \int_0^D \sigma(x)\,dx \tag{3.5}$$

where n_M = number of analyte nuclides per g; i = beam intensity; λ_M = decay constant; t_b = irradiation time; $D = \rho L$, with ρ = density of the monitor foil, L = thickness.

Therefore, when a standard (S) and a sample (X) are irradiated separately, both placed behind a monitor foil with the same thickness and composition

$$\frac{i_X}{i_S} = \frac{a_{M,X}}{(1 - e^{-\lambda_M t_{b,X}})} \frac{(1 - e^{-\lambda_M t_{b,S}})}{a_{M,S}} \tag{3.6}$$

where $a_{M,X}$, $a_{M,S}$ = count-rates of the monitor foil at the end of the irradiation, for the sample and the standard, respectively; $t_{b,X}$, $t_{b,S}$ = irradiation times of the sample and the standard.

When choosing a monitor foil the following factors must be taken into account.

(1) The thickness must be small compared with the range.
(2) The foil must have a constant thickness.
(3) The material must have a high thermal conductivity, so metal foils are usually preferred.
(4) Activity measurements are easiest when only a limited number of γ-ray emitters, with a simple γ-ray spectrum, are produced.

Disadvantages of the use of monitor foils are as follows.

(1) Additional activity measurements are required.
(2) Energy degradation in the monitor foil introduces an additional uncertainty in the incident energy value.
(3) Radionuclides from the monitor foil recoil into the sample.

The last disadvantage is, in general, negligible for the standards, where the induced activity is high compared to the contamination. For samples with a low concentration of the analyte element, and especially when pure β^+-emitters are to be measured, spectral interference may occur from the recoil effect. Chemical etching of the sample surface after the irradiation generally allows this source of error to be avoided. To minimize the influence of recoil nuclei when analysing metal samples, a monitor foil of the same composition as the sample may be used. Often, an additional foil is also placed between the monitor foil and the sample. This additional foil may have the same composition as the (metal) sample or may be made of a material that is activated to only a negligible extent.

Table 3.2 gives some characteristics of metal foils useful for beam intensity monitoring.

3.2   MEASUREMENT OF THE INDUCED ACTIVITY

3.2.1   Measurements

A large number of radionuclides formed by irradiation with charged particles are positron (β^+) emitters. The energy spectrum of the positrons is continuous, with a maximum energy characteristic of the radionuclide. The β-rays emitted can be measured directly, for instance with plastic scintillators or GM-counters. When the sample is not infinitely thin, attenuation must be taken into account. Furthermore, the determination of the maximum energy is not very selective.

Therefore, when, in addition to β-particles, γ-rays are emitted, γ-ray detection is preferred. Gamma-ray detection with NaI(Tl) detectors has, except for the detection of pure positron emitters (see below), almost disappeared from activation analysis. Because of their superior resolution [more than one order of magnitude better than that of an NaI(Tl) detector] germanium semiconductor detectors are preferable. To date, these detectors are made from high-purity germanium (HPGe) and have a resolution, i.e. the full peak width at half-maximum (FWHM) for the 1332-keV line of ^{60}Co, of less than 2.0 keV, and an efficiency, i.e. the detection efficiency relative to a 7.5×7.5 cm NaI(Tl) scintillation detector for the 662-keV line of ^{137}Cs, of up to 25%.

The principles of γ-ray spectrometry are described in references [9–11].

When the radionuclide formed emits positrons, but no γ-rays, as for instance ^{15}O, ^{11}C, ^{13}N and ^{18}F, it is usually preferable to measure the annihilation radiation. When a positron interacts with an electron, two γ-rays with an energy of 511 keV are emitted at an angle of 180° to each other (annihilation radiation). Measurement of annihilation photons requires that the annihilation is complete, which means that the sample is surrounded with a sufficient amount of material for all the β^+-radiation emitted to be completely absorbed and yield 511 keV radiation.

Since the energy of the annihilation radiation does not yield any qualitative information, NaI(Tl) detectors are usually used. These are available in large sizes, e.g. 7.5 cm diameter by 7.5 cm height, and thus have a high detection efficiency, and they are less expensive than germanium semiconductor detectors. The superior energy resolution of the latter is in this case not a significant advantage. In order to detect the annihilation radiation more selectively, two NaI(Tl) detectors on opposite sides of the sample and in coincidence are often used.

Figure 3.10 shows a typical experimental set-up, in use at the Institute of Nuclear Sciences of Ghent University. It consists of two cylindrical NaI(Tl) detectors with a diameter of 7.5 cm and a height of 7.5 cm, placed on opposite sides of the sample and in line with it, in a 6-cm thick lead shield. The photomultiplier tubes of the detectors are each coupled by a preamplifier to a linear amplifier. The output signals of these amplifiers (A, B) are sent to a timing single-channel analyser (SCA). When the height of the input pulse exceeds the lower threshold and is below the higher threshold, the analyser generates a logical output pulse (C, D). The threshold settings correspond to 400 and 600 keV. The output pulse (C, D) has a presettable

Table 3.2 — Characteristics of metal foils useful for beam intensity monitoring.

Particle	Monitor foil	Melting point, °C	Thermal conductivity at 25°C, $W.cm^{-1}.°C^{-1}$	Activation reaction	Threshold energy, MeV	Product half-life	Main γ-rays, keV
p	Cu	1083	3.98	$^{63}Cu(p,n)^{63}Zn$	4.2	38.5 min	679
				$^{65}Cu(p,n)^{65}Zn$	2.1	243.7 d	1115
	Ti	1675	0.2	$^{48}Ti(p,n)^{48}V$	4.9	16 d	984, 1312
	Zr	1852	—	$^{92}Zr(p,n)^{92m}Nb$	2.8	10.1 d	935
d	Ni	1453	0.90	$^{60}Ni(d,n)^{61}Cu$	$Q > 0$	3.37 hr	283, 656
	Ti	1675	0.27	$^{47}Ti(d,n)^{48}V$	$Q > 0$	16 d	984, 1312
	Mo	2610	1.4	$^{92}Mo(d,n)^{93m}Tc$	$Q > 0$	43 min	390
^{3}He	Cu	1083	3.98	$^{65}Cu(^3He,2n)^{66}Ga$	5.0	9.5 hr	1039
α	Cu	1083	3.98	$^{63}Cu(\alpha,n)^{66}Ga$	8.0	9.5 hr	1039
	Ni	1453	0.90	$^{60}Ni(\alpha,2n)^{62}Zn$	18.2	9.3 hr	548, 597

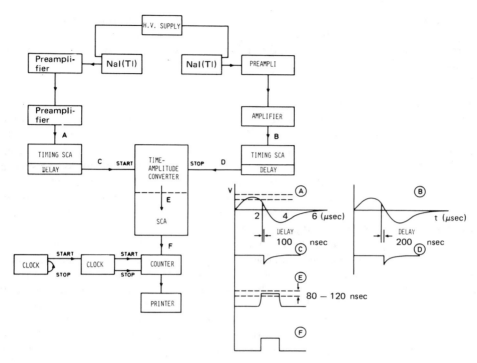

Fig. 3.10 — Set-up for γ-γ coincidence measurement of annihilation radiation.

delay relative to the moment when the decreasing input signal (A, B) reaches 50% of its maximum amplitude (constant fraction discrimination). The output pulse in the right-hand channel (D) is delayed 100 nsec more than the one in the left-hand channel (C). The pulses C and D are sent to the start and stop input of a time-to-amplitude converter/single channel analyser. This module gives a linear signal (E) with an amplitude proportional to the time difference between the start and stop pulses, which is internally sent to the single-channel analyser. The single-channel analyser generates a logical signal (F) for each input pulse within the preset limits. For a pure positron emitter the frequency distribution of the pulse heights from the time-to-amplitude converter has a maximum at around 100 nsec. The window of the single channel analyser is centred around this maximum, with a 40 nsec width. Thus 96% of the pulses from the time-to-amplitude converter are recorded for a pure positron emitter. The logical ouput signals from the single-channel analyser are counted with a scaler controlled by two clocks, thus allowing cyclic measurements: one clock measures and controls the period (measuring time + waiting time between two successive counts), the other the measuring time. During the measuring time a printer prints the content of the scaler. The natural background is ca. 0.01 counts/sec. Disc-shaped samples are placed in a cylindrical sample holder and measured in a fixed geometry between the two detectors. Solutions are measured with the same geometry. The ratio of the detection-efficiency for a disc-shaped source (e.g. the standard) to that of a solution (e.g. the sample) can be determined by measuring a copper foil irradiated with thermal neutrons (^{64}Cu), in the same geometry as the standard. The foil is then dissolved in $6M$ nitric acid, the volume adjusted and the measurement repeated.

3.2.2 Analysis of decay curves

The measurement of positron emitters through annihilation radiation is not specific, since a large number of radionuclides produced by irradiation with charged particles emit positrons. Since the energy of the annihilation radiation does not yield qualitative information, the half-life is used as a qualitative criterion. When several positron emitters occur simultaneously, the decay curve must be analysed.

A decay curve, i.e. the plot of the activity as a function of time, can be analysed graphically. This method is, however, subjective and time-consuming, so in most cases a suitable computer program is used.

Several programs with a varying degree of complexity have been described, that by Cumming [12] being most often used. This program calculates, by the method of least squares, the initial count-rate of each component of a composite decay-curve, if the number of components and their half-lives are known. Starting from an approximate value for the half-life of one or several components the best half-life for each component can be deduced. The following data must be entered: number of measurement points, number of components, components for which a best half-life must be calculated, background count-rate, and approximate half-life for all components. For each measurement point the starting time, the measuring time and the number of counts is entered. In the first step the calculations are done with the half-lives given. The count rate of each component at the start of the first measurement and the standard deviation are obtained. In the second step, an iterative procedure is applied to calculate the best half-life for one or several components. The half-life, with its standard deviation, as well as the initial count-rates (with their standard deviations) are obtained. To verify that the fitted curve is in good agreement with the experimental one, the χ^2-test is applied.

An experimentally determined decay curve consists of n measurements of the count-rate a_i at time t_i ($i = 1, \ldots n$). If m independent radionuclides with starting activity ($t = 0$) $a_{0,j}$ and disintegration constant λ_j are present, the count-rates are given by n equations of the form:

$$a_i = \sum_{j=1}^{m} a_{0,j} e^{-\lambda_j t_i} + v_i \tag{3.7}$$

Each term is the contribution of one component. The difference v_i between the measured and expected count-rate is due to statistical fluctuations. Since the m coefficients $a_{0,j}$ occur linearly in Eq. (3.7), they can be calculated by using the method of weighted least squares. The sum

$$\sum_{i=1}^{n} \frac{v_i^2}{s_i^2} = \sum_{i=1}^{n} \frac{1}{s_i^2} \left[a_i - \sum_{j=1}^{m} a_{0,j} e^{-\lambda_j t_i} \right]^2 \tag{3.8}$$

must be minimized. When the background count-rate B is deduced from a measurement during time $t_{m,B}$, and the count-rate a_i from a measurement during $t_{m,i}$, it can easily be shown that s_i^2, the variance of a_i, is given by

$$s_i^2 = \frac{a_i + B}{t_{m,i}} + \frac{B}{t_{m,B}} \tag{3.9}$$

$\sum_{i=1}^{n} \dfrac{v_i^2}{s_i^2}$ is minimal if

$$\frac{\delta}{\delta a_{0,k}} \sum_{i=1}^{n} \frac{v_i^2}{s_i^2} = 0 \tag{3.10}$$

or

$$\sum_{i=1}^{n} \frac{1}{s_i^2} \left(a_i - \sum_{j=1}^{m} a_{0,j} e^{-l_j t_i} \right) e^{-\lambda_j t_i} = 0 \tag{3.11}$$

with $k = 1, 2 \ldots m$. This set of m equations is solved for the m unknowns $a_{0,j}$ by Crout's method [13].

Since the λ_j-values do not occur linearly in Eq. (3.7), the best value for λ_j cannot be calculated directly by the method of least squares. λ_j' and $a_{0,j}'$ are approximate values of λ_j and $a_{0,j}$, such that

$$\lambda_j = \lambda_j' + \Delta\lambda_j \tag{3.12}$$

$$a_{0,j} = a_{0,j}' + \Delta a_{0,j} \tag{3.13}$$

$a_{0,j}'$ is obtained in the first calculation step, where λ_j is considered known and equal to λ_j'. For the q components for which a best half-life must be calculated,

$$a_{0,j} e^{-\lambda_j t_i} = (a_{0,j}' + \Delta a_{0,j}) e^{-(\lambda_j' + \Delta\lambda_j) t_i} \tag{3.14}$$

If $\Delta\lambda_j t_i$ is small

$$e^{-\Delta\lambda_j t_i} \sim 1 - \Delta\lambda_j t_i \tag{3.15}$$

Equation (3.14) then becomes

$$a_{0,j} e^{-\lambda_j t_i} \sim a_{0,j} e^{-\lambda_j t_i} - a'_{0,j} \Delta\lambda_j t_i e^{-\lambda'_j t_i} \tag{3.16}$$

the term $-\Delta a_{0,j}\Delta\lambda_j t_i e^{-\lambda_j t_i}$ being neglected.

Combination of Eqs. (3.16) and (3.8) and calculation of the derivative with respect to $a_{0,j}$ and $a'_{0,j}\Delta\lambda_j$ yields $m + q$ equations with $m + q$ unknowns ($a_{0,j}$ and $a'_{0,j}\Delta\lambda_j$). Dividing $a'_{0,j}\Delta\lambda_j$ by $a'_{0,j}$ yields $\Delta\lambda_j$. The approximate value λ'_j is then corrected by adding $\Delta\lambda_j$ and the resulting value is used as λ'_j in the following calculation step along with the $a_{0,j}$, which is used as $a'_{0,j}$. This iterative procedure is repeated until $\Delta\lambda_j$ is sufficiently small. When the original λ'_j is sufficiently close to the real value, so that Eq. (3.12) is a good approximation, the successive λ_j values will converge.

3.3   TREATMENT OF THE SAMPLE SURFACE AFTER THE IRRADIATION

One of the main advantages of activation analysis is that after the irradiation the activity formed at the sample surface can, for compact samples such as pieces of metals and semiconductors, be removed, so that after the irradiation no contamination with non-radioactive material need be feared. Usually, and certainly when carbon, nitrogen and oxygen are determined, a surface layer is removed after the irradiation. In this way radionuclides formed from the analyte element at the sample surface and that have penetrated into the sample because of nuclear recoil, as well as radionuclides formed in the monitor foil that have recoiled into the sample, are removed. The analyte element may be present in the oxide layer covering most metals, as is the case for carbon, nitrogen and oxygen. In addition, it may originate from accidental contamination of the sample before the irradiation.

3.3.1   Recoil of radionuclides

The thickness of the surface layer that must be removed after the irradiation depends on the range of the recoil nuclei.

The energy of the recoil nucleus for a given nuclear reaction can be calculated as a function of the angle of emission, by means of the equations given by Marion and Young [14]. For the nuclear reaction A(a, b)B, which is schematically shown in Fig. 3.11, the energy of the heavy product nucleus B in the laboratory system is given by Eq. (3.17).

$$E_B = (E_a + Q)\alpha \left[\cos\theta \pm \left(\frac{\gamma}{\alpha} - \sin^2\theta \right)^{1/2} \right]^2 \tag{3.17}$$

where

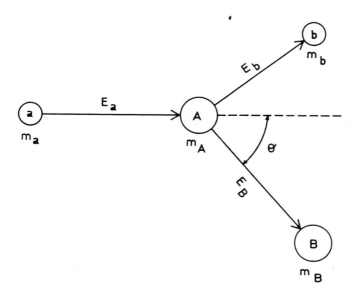

Fig. 3.11 — Kinematics of nuclear reactions.

$$\alpha = \frac{m_a m_B E_a}{(m_a + m_A)(m_b + m_B)(E_a + Q)} \tag{3.18}$$

$$\gamma = \frac{m_A m_b}{(m_a + m_A)(m_b + m_B)}\left[1 + \frac{m_a}{m_A}\frac{Q}{(E_a + Q)}\right] \tag{3.19}$$

In Eq. (3.17) only the plus sign must be used, unless $\alpha < \gamma$. In that case:

$$\theta_{max} = \sin^{-1}(\gamma/\alpha)^{1/2} \tag{3.20}$$

As an example of the use of these equations, the recoil energy of ^{18}F from the $^{16}O(^3He, p)^{18}F$ reaction ($Q = 2.0\,MeV$) may be calculated for different incident energies. For an incident energy of 18 MeV, α and γ are equal to 0.1346 and 0.04515, respectively; α is thus larger than γ, so for every angle of emission smaller than θ_{max} ^{18}F nuclei with two different energies can be observed, corresponding to the plus and minus signs in Eq. (3.17). The energy of the ^{18}F nuclei emitted in the forward direction ($\theta = 0$) amounts to 6.7 and 0.48 MeV, and the maximum angle of emission, θ_{max}, is 35.4°. Figure 3.12 gives the ^{18}F energy as a function of θ for incident energies of 18, 16 and 14 MeV.

The stopping power for the recoil nuclei of different elements is given in the compilation by Ziegler [15]. The range can be calculated by using Eq. (2.17). Ranges of heavy ions are also given in the compilation by Northcliffe and Schilling [16]. For

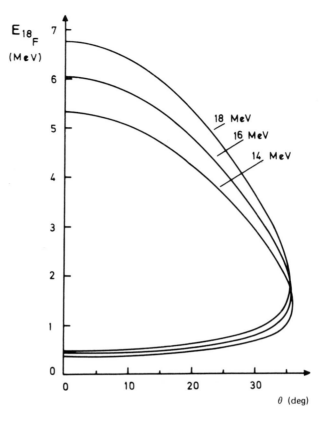

Fig. 3.12 — ^{18}F-energy [reaction: ^{16}O(^{3}He, p)^{18}F] as a function of the angle of emission.

6.7-MeV ^{18}F ions the range is 1.2 mg/cm^2 (4.4 μm) in aluminium, 2.1 mg/cm^2 (2.4 μm) in nickel, and 4.6 mg/cm^2 (2.8 μm) in tantalum. Since the compilation by Northcliffe and Schilling [16] is not very accurate for < 1 MeV per nucleon [17], it is recommended to remove a surface layer thickness that is several times that calculated.

3.3.2 Removal of a surface layer after the irradiation
After the irradiation a surface layer can be removed from massive samples by:

— chemical etching;
— mechanical grinding;
— a combination of both.

Mechanical grinding alone is not often applied, since it is difficult to avoid contamination of the deeper layers of the sample. In addition, for disc-shaped samples with an irradiated face which is not completely flat or with faces which are not strictly parallel, systematic errors may occur, because more material is removed from the thicker part of the sample.

Chemical etching alone has been applied by various authors [18–20]. When choosing an appropriate etching solution, usually a mixture of acids, several factors must be considered:

— speed: the time required for etching must be short compared to the half-life of the radionuclide considered;
— smoothness of the surface obtained: this may be checked by microscopy or using a surface-texture measuring instrument;
— reproducibility of the thickness removed.

These requirements are sometimes contradictory. When possible, etching is done at room temperature, but sometimes a constant higher temperature is preferred. It is recommended to etch at least twice, in two etching solutions with the same composition. Afterwards the sample is rinsed with water, then methanol or acetone, and dried.

Some authors recommend a combination of mechanical grinding and chemical etching [21, 22]. Blondiaux et al. [23, 24] showed that under given irradiation conditions a 'carbon layer' is deposited on the sample surface at the beam spot. A mechanism for the formation of this layer was proposed by Blondiaux [25] and it was shown that the formation of the 'carbon layer' takes place only in the irradiated zone. The velocity of formation of the 'carbon layer' is determined by the following factors.

(1) The current density and the temperature of the target. The curve that gives the velocity of growth of the 'carbon layer' as a function of the current density shows a maximum at around 150–200 nA/mm². For temperatures above 100°C the velocity is very low and at still higher temperatures desorption of the carbon-containing products takes place until a given surface concentration, depending on the temperature, is reached.
(2) The energy and nature of the incident ion. The energy has only a small influence on the velocity of growth, but the nature of the incident ion has an important influence.
(3) The nature of the target.
(4) The quality of the vacuum.

The 'carbon layer' delays chemical etching of the sample after the irradiation. Figure 3.13 gives the aspect of the sample surface after irradiation and chemical etching, as determined by using a surface-texture measuring instrument. It is clear that the sample is etched to a lesser extent inside the beam spot than outside. If a light mechanical polishing is applied after the irradiation and before chemical etching, normal etching takes place. Mechanical polishing was done with felt impregnated with very fine diamond powder (0.2 μm). For germanium, less than a 40-nm thickness was removed in this way. It was also shown [23, 24] that incomplete etching after the irradiation may cause significant errors for the determination of oxgen in

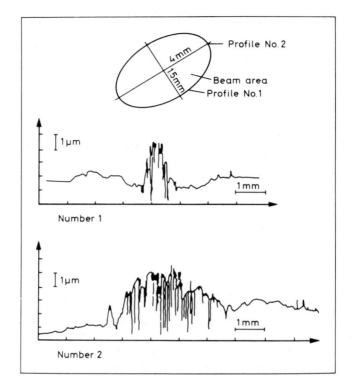

Fig. 3.13 — Surface profiles for a germanium sample irradiated with 3 MeV tritons and chemically etched. The intensity of the beam was 6 μA/cm^2. (Reproduced with permission, from G. Blondiaux, A. Giovagnoli, K. Ishii, C. Koemmerer and M. Valladon, *J. Radioanal. Chem.*, 1980, **55**, 407. Copyright 1980, Akadémiai Kiadó, Budapest).

germanium by use of the $^{16}O(t, n)^{18}F$ reaction, when only a chemical etching is used. When the sample was mechanically polished before chemical etching, no ^{18}F was detected in the germanium, even when it had been irradiated behind a mica foil.

3.3.3 Measurement of the thickness removed
In charged particle activation analysis the thickness of the layer removed must be known, since it determines the energy effectively incident on the sample.

Several methods can be applied in order to determine the thickness of the removed layer.

(1) Measurement of the thickness with a mechanical micrometer before and after etching.
(2) Measurement of the thickness with an electronic micrometer before and after etching.
(3) Weighing the sample before and after the etch.
(4) Covering the non-irradiated side of the sample before the etch and measuring

the thickness within the irradiated area by means of an electronic micrometer
before and after etching.
(5) Measurement of the activity formed from the matrix, before and after etching.

For methods (1)–(3) the thickness of the sample is in general measured before the
irradiation, instead of before the chemical etching, in order to avoid contamination
of the measuring equipment and for reasons of radiation protection.
    Method (1) is very simple, but the accuracy is seldom better than a few μm. In
addition, the surface of the measuring instrument is large, so for rough surfaces,
sometimes obtained after etching, the maximal thickness is determined instead of the
average thickness.
    Method (2) makes use of an electronic micrometer (e.g. Tesatronic digital, Tesa,
Renens, Switzerland). The experimental set-up is shown in Fig. 3.14. The two bullet-

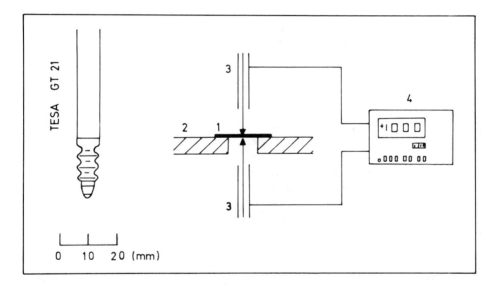

Fig. 3.14 — Electronic micrometer: measuring probe TESA GT21 and experimental set-up: 1,
sample; 2, measuring table; 3, measuring probes; 4, digital measuring module.

shaped measuring probes have a 3 mm diameter. The accuracy obtainable is
$\pm 0.1\ \mu$m for a thickness below 0.2 μm and the thickness must be measured at several
spots, before and after the etching, to obtain the average thickness removed.
    Weighing the sample [method (3)] is simple and very convenient, when done with
a digital electronic balance. For this method the sample density need not be known,
since the thickness removed is directly obtained in mg/cm^2, as required for calculating the effective incident energy, and an average value for the thickness removed is
directly obtained. Methods (1)–(3) assume that the irradiated and non-irradiated
parts of the sample are etched at the same rate. Method (2) allows verification of this

assumption, if the beam diameter is smaller than the diameter of the sample and measurements in the irradiated area are compared with measurements outside it.

To avoid systematic errors due to a different etching speed inside and outside the irradiated area, Valladon and Debrun [22] proposed a method for measuring the thickness removed within the irradiated area. After the irradiation the non-irradiated side of the sample is glued onto a graphite block and the electronic micrometer is set to zero with the probe on the irradiated area. The sample is then etched, still attached to the graphite block, placed back under the measuring instrument in exactly the same position and again measured within the irradiated area. A similar method was described by Sanni *et al.* [20]. Some disadvantages are:

— the irradiated spot must be large compared to the tip of the measuring probe;
— the accuracy depends on the roughness of the surface after etching;
— the density must be known.

Method (5), proposed by Valladon *et al.* [26] is based on the measurement of the activity produced from the matrix. It follows from Eq. (2.46) that

$$Y(E) = \frac{a}{i[1 - e^{-\lambda t_b}]} = k \int_E^{E_T} \frac{\sigma(E)dE}{dE/dx(E)} \qquad (3.21)$$

where $Y(E)$ = normalized thick target yield. For the matrix under consideration Y is determined experimentally as a function of the energy. After irradiation of the sample and chemical etching, the normalized thick target yield $Y(E_r)$ is measured in the same experimental conditions as for the measurement of $Y = Y(E)$. As shown in Fig. 3.15, the residual energy after etching, E_r, can be derived from the $Y = Y(E)$ curve. The thick target yield before and after the etching can also be measured, to give the ratio

$$R = \frac{Y(E_r)}{Y(E_I)} \qquad (3.22)$$

where E_I = incident energy before etching. E_r is then deduced from the curve giving R as a function of the energy. Advantages of this method are that relative yields determined in any measuring geometry may be used and that it is not necessary to refer to the conditions used when determining the thick target yield curve. A disadvantage is, of course, that two activity measurements are required. Method (5) has the following advantages:

— no special equipment is needed;
— the density need not be known;
— the value obtained for E_r represents an average, taking into account possible surface roughness, which is important as smooth surfaces are not always obtained after chemical etching.

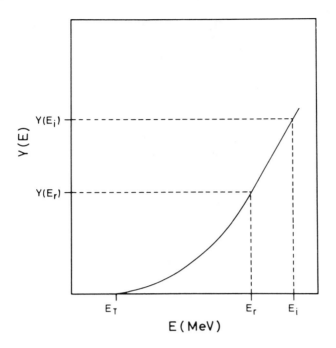

Fig. 3.15 — Normalized thick target yield as a function of the energy. (Reproduced by permission, from M. Valladon, G. Blondiaux, A. Giovagnoli, C. Koemmerer and J. L. Debrun, *Anal. Chim. Acta*, 1980, **116**, 25. Copyright 1980, Elsevier Science Publishers).

Disadvantages are:

— to obtain good precision, the incident energy must be chosen in a range where $Y(E)$ varies rapidly with the energy;
— a suitable nuclear reaction with the matrix, yielding a radionuclide with suitable decay characteristics must be available.

3.4 CALCULATIONS

3.4.1 Incident energy

The monitor foil, which is placed before the sample during the irradiation, the additional foils, placed between the monitor foil and the sample to stop recoil nuclei, and the layer removed by chemical etching after the irradiation, degrade the energy of the incident particles. The monitor foil and the additional foils placed before the standard also degrade the energy of the incident particles.

The effective incident energy, i.e. the energy incident on the part of the sample which is actually measured, is calculated by means of appropriate range–energy tables. When a beam with energy E_I impinges on a target with thickness L, the energy of the particles emitted, E_U, is calculated as follows. From the range–energy

calculated for instance by linear interpolation. When the energy values are chosen close enough to each other, linear interpolation is sufficiently accurate. $R(E_U)$ is given by Eq. (3.23):

$$R(E_U) = R(E_I) - \rho L \tag{3.23}$$

where ρ is the density of the target.

By linear interpolation, E_U is deduced from the range–energy table. An example is given in Table 3.3. A gold sample, placed behind a 9.35 mg/cm² molybdenum

Table 3.3 — Calculation of incident energy.

Molybdenum R, mg/cm²	E_d, MeV	Gold R, mg/cm²
53.66 ←	5.00	
↓		
−9.35		
↓		
44.31 ⟶	4.42 ⟶	70.13
		↓
		−9.80
		↓
	3.98 ←	60.33

monitor foil is irradiated with 5 MeV deuterons. After the irradiation a 9.80 mg/cm² surface layer is removed by etching. The energy after traverse of the monitor foil is 4.42 MeV, after the layer removed by etching it is 3.98 MeV, the effective incident energy.

3.4.2 Activity of the standard

The basic equation of charged particle activation analysis, Eq. (2.49), assumes that the incident energy is the same for the sample and the standard. In practice, this is sometimes difficult to achieve in cases where the sample is chemically etched after the irradiation, since the chemical etch and thus the effective incident energy are not always exactly reproducible. Therefore, several standards are irradiated at different incident energies. For the determination of calcium in cast iron [27] by the $^{40}Ca(\alpha, p)^{43}Sc$ reaction (Table 3.4), for instance, three series of calcium carbonate standards are irradiated, with 14.40, 13.94 and 13.46 MeV α-particles, respectively. This allows the determination of $a_S/i_S (1 - e^{-\lambda t_{b,S}})$ as a function of the energy. When a thickness between 3.24 and 7.10 mg/cm² is removed from the sample by chemical etching, the effective incident energy is between 14.40 and 13.46 MeV. The corresponding $a_S/i_S(1 - e^{-\lambda t_{b,S}})$ to be used in Eq. (2.49) may be deduced from the curve

Table 3.4 — Standardization curve for calcium in cast iron [27], determined by the $^{40}Ca(a, p)^{43}$ reaction. The standard is calcium carbonate, the incident energy is 17 MeV.

	Monitor foil, mg/cm²	Additional foils, mg/cm²	Energy, MeV	To be removed by etching*, mg/cm²
Series 1	Cu 6.90	Al 4.15	14.40	3.24
Series 2	Cu 6.90	Al 5.72	13.94	5.17
Series 3	Cu 6.90	Al 7.28	13.46	7.10

* The sample is placed behind a copper foil (6.90 mg/cm²) and an aluminium foil (1.56 mg/cm²).

that gives $a_S/i_S(1 - e^{-\lambda t_{b,s}})$ as a function of the energy, the standardization curve. Figure 3.16 shows the standardization curve for calcium in cast iron.

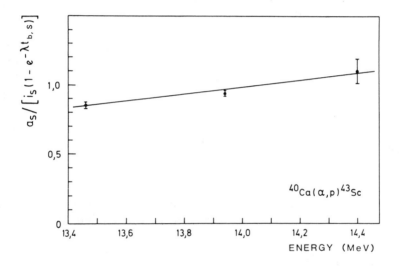

Fig. 3.16 — Standardization curve for the determination of calcium in cast iron [27].

3.4.3 Correction factor $F(E_I)$
$F(E_I)$, defined by Eq. (2.53), can be calculated by numerical integration. From the excitation function an energy–cross-section table is deduced, consisting of n pairs E_i–$\sigma(E_i)$, with $\sigma(E_1) = 0$. The numerator and the denominator $[INT(E_i)]$ of the correction factor are calculated for the $n - 1$ energies E_i $(i = 2, \ldots n)$ by means of Eq. (3.24):

$$INT(E_i) = \int_{E_1}^{E_i} \frac{\sigma(E)dE}{S(E)}$$

$$\sim \frac{1}{2}\sum_{j=2}^{i}\left[\frac{\sigma(E_j)}{S(E_j)}+\frac{\sigma(E_{j-1})}{S(E_{j-1})}\right](E_j-E_{j-1}) \tag{3.24}$$

Table 3.5 gives the results of the calculations for the determination of oxygen in

Table 3.5 — $F(E_1)$ for the determination of oxygen in zirconium by using the $^{16}O(^3He, p)^{18}F$ reaction. The standard is quartz [28].

E_i	$\sigma(E_i)$	$INT(E_i)_S$	$INT(E_i)_X$	$F(E_i)$
3.0	0			
4.0	20	0.015	0.027	0.559
5.0	80	0.103	0.182	0.566
6.0	180	0.358	0.625	0.573
7.0	340	0.925	1.59	0.579
8.0	380	1.78	3.04	0.585
9.0	350	2.72	4.61	0.589
10.0	280	3.60	6.06	0.593
11.0	250	4.40	7.37	0.596
12.0	220	5.15	8.60	0.599
13.0	180	5.84	9.70	0.601
14.0	160	6.46	10.70	0.604
15.0	130	7.01	11.58	0.605
16.0	120	7.52	12.38	0.607
17.0	100	7.98	13.11	0.609
18.0	90	8.40	13.76	0.610
19.0	80	8.79	14.37	0.612

zirconium by using the $^{16}O(^3He, p)^{18}F$ reaction [28] and quartz as a standard. It appears that $F(E_1)$ varies slowly with the energy, so for intermediate energies linear interpolation can be applied.

3.4.4 Concentration
The concentration is calculated by means of Eq. (3.25):

$$c_X = c_S \frac{a_X}{i_X(1-e^{-\lambda t_{b,x}})}\frac{i_S(1-e^{-\lambda t_{b,s}})}{a_S}F(E_1) \tag{3.25}$$

where c_X, c_S = concentrations of the analyte element in the sample and the standard; a_X = count-rate for the sample at the end of the irradiation, deduced from the count-rate $a_{0,j}$ at the start of the first measurement (decay curve analysis) at time t_w after irradiation, by means of Eq. (3.26):

$$a_X = a_{0,j}e^{\lambda t_w} \tag{3.26}$$

or deduced from the number of counts during one measurement (germanium semiconductor) by using Eq. (2.50); a_S = count-rate for the standard at the end of the irradiation; i_X/i_S = ratio of the beam intensities for the sample and the standard (Section 3.1.3); $F(E_I)$ = correction factor at energy E_I.

3.5 SOURCES OF ERROR

In addition to random errors due, for instance, to statistical uncertainties in the detected activities, systematic errors may occur because of the use of inaccurate data for the calculation of the concentration. Examples of data that may be inaccurate are:

(1) beam energy;
(2) thickness of the monitor foil and of the additional foils;
(3) thickness of the layer removed by chemical etching (see Section 3.3.3);
(4) stopping power;
(5) excitation function;
(6) half-life of the radionuclide produced;
(7) isotopic abundance of the radionuclide considered.

An inaccurate beam energy causes a systematic error in the calculation of $F(E_I)$ and the calculation of the energy degradation. Since $F(E_I)$ varies only slowly with energy (Section 3.4.3), the error in $F(E_I)$ as a result of an inaccurate beam energy is in general negligible. When metal samples are analysed, and are chemically etched after the irradiation, an inaccurate beam energy causes an error in the thickness that must be removed by chemical etching. Assume that the energy after traverse of the monitor foil and of the additional foils placed before the sample and the standard is E_0, and that l_X is the thickness removed from the sample by chemical etching and l_F is the corresponding thickness of the energy degrading foils placed before the standard to degrade the energy to E_I; l_X and l_F are related by Eq. (3.27):

$$l_F \rho_F S_F = l_X \rho_X S_X \tag{3.27}$$

where S_F, S_X = stopping powers at $(E_0 + E_I)/2$ for the energy degrading foils and for the sample. At the exact incident energy, $E_0 + \Delta E_0$ instead of E_0, l_F corresponds to $l_X + \Delta l_X$, and to $E_I + \Delta E_I$:

$$l_F \rho_F S_F' = (l_X + \Delta l_X) \rho_X S_X' \tag{3.28}$$

where S_F' and S_X' = stopping powers at $(E_0 + \Delta E_0 + E_I + \Delta E_I)/2$. It follows from Eqs. (3.27) and (3.28) that

$$\frac{l_X}{l_X + \Delta l_X} = \frac{S_F}{S_F'} \frac{S_X'}{S_X} \tag{3.29}$$

It was shown in Section 2.3.4 that S_X/S_F varies only slightly with the energy, except for very low energy. Therefore $l_X + \Delta l_X$ is to a first approximation equal to l_X, and the resulting error is negligible.

An inaccurate thickness of the monitor foil or of an additional foil placed before the sample and the standard also results in an inaccurate E_0, and can be treated in the same way.

An error Δl_X in the thickness of the layer removed by chemical etching or Δl_F in the thickness of a foil that is placed only before the standard, results in an error $\Delta l_X \rho_X S_X$ or $\Delta l_F \rho_F S_F$ in the incident energy, so the incident energies for the standard and the sample are different. The error in the concentration can be deduced from the standardization curve, i.e. the curve that gives $a_S/i_S(1 - e^{-\lambda t_{b,S}})$ as a function of the energy. For the same $\rho \Delta L$, the error increases with S and with the slope of the standardization curve at E_I. It is thus advantageous to choose an energy where the standardization curve has only a small slope.

Thus far, it was assumed that the foils placed before the sample and the standard are homogeneous in thickness. The monitor foils placed before the sample and the standard thus have exactly the same thickness and cause the same energy degradation. If this is not the case, a systematic error occurs that may be estimated in the same way as for an inaccurate thickness of the etched layer. The occurrence of this source of error can be detected by irradiating a number of standards under the same conditions, in order to test the reproducibility of $a_S/i_S(1 - e^{-\lambda t_{b,S}})$.

In general, the influence of an inaccurate incident energy is small when the same error occurs for the sample and the standard. When the error occurs only for the standard or for the sample, the resulting systematic error may be more important.

The influence of sources of error Nos. (4) and (5) has been treated in detail in Sections 2.3.2 and 2.3.3.

The uncertainty in the half-life of the radionuclides commonly used in CPAA is in general rather small (0.1–1%). This uncertainty can lead to an error in the final analytical result obtained by calculation of the count-rate at the end of the irradiation by using Eq. (2.50) and the calculation of the saturation factors in Eq. (2.49).

When the isotopic abundance of the nuclide that gives the nuclear reaction considered is different in the sample and the standard, a systematic error occurs. In certain samples the isotopic abundance is indeed significantly different from the accepted value (Commission on Atomic Weights, IUPAC, 1971). In chemical reagents an isotopic abundance of 3.75% was observed for ^6Li, whereas the accepted value is 7.5%. The natural isotopic abundance of ^{10}B ranges from 19.8 to 20.1% [29]. Another element for which important variations of the isotopic abundance may occur is lead.

As explained in Section 2.3, it is usually assumed that the beam intensity is constant during the irradiation. If this is not the case, systematic errors are made. The assumption can be validated by monitoring the current incident on the sample (Section 3.1.3). Usually the operator of, for example, a cyclotron keeps the beam intensity constant within a few per cent. The systematic error made is then negligible, if the irradiation time is not long compared to the half-life of the radionuclide of interest.

REFERENCES

[1] M. Livingston and J. Blewet, *Particle Accelerators*, McGraw-Hill, New York, 1962.

[2] K. W. Allen, in *Nuclear Spectroscopy and Reactions*, Part A, J. Cerny (ed.), p. 3. Academic Press, New York, 1974.

[3] J. Livingood, *Principles of Cyclic Particle Accelerators*, Van Nostrand, Princeton, 1961.

[4] B. G. Harvey, in *Nuclear Spectroscopy and Reactions*, Part A, J. Cerny (ed.), p. 35. Academic Press, New York, 1974.

[5] C. Vandecasteele, *Ph. D. Thesis*, Ghent University, Belgium, 1975.

[6] R. Mortier, *Ph. D. Thesis*, Ghent University, Belgium, 1983.

[7] R. Mortier, C. Vandecasteele, J. Hertogen and J. Hoste, *J. Radioanal. Chem.* 1982, **71**, 189.

[8] G. Wauters, C. Vandecasteele and J. Hoste, *J. Radioanal. Nucl. Chem., Articles*, 1968, **98**, 345.

[9] C. E. Crouthamel, F. Adams and R. Dams, *Applied Gamma-Ray Spectrometry*, Pergamon Press, Oxford, 1970.

[10] G. F. Knoll, *Radiation Detection and Measurement*, Wiley, New York, 1979.

[11] U. Herpers, in *Treatise on Analytical Chemistry*, Part I, 2nd Ed., Vol. 14, P. J. Elving (ed.), p. 123. Wiley-Interscience, New York, 1986.

[12] J. Cumming, *Applications of Computers to Nuclear and Radiochemistry*, G. O'Kelley, (ed.), *NAS-NS* 3107, 1963.

[13] P. D. Crout, *Trans. A.I.E.E.* 1941, **60**, 1235.

[14] J. B. Marion and F. C. Young, *Nuclear Reaction Analysis, Graphs and Tables*, North-Holland, Amsterdam, 1968.

[15] J. F. Ziegler, *Handbook of Stopping Cross-Sections for Energetic Ions in all Elements*, Pergamon Press, New York, 1980.

[16] L. C. Northcliffe and R. F. Schilling, *Nuclear Data Tables*, 1970, **A7**, 253.

[17] J. S. Forster, D. Ward, H. R. Andrews, G. C. Ball, G. J. Costa, W. G. Davies and I. V. Mitchell, *Nucl. Instr. Meth.*, 1976, **136**, 349.

[18] C. Vandecasteele, F. Adams and J. Hoste, *Anal. Chim. Acta*, 1973, **66**, 27.

[19] K. Strijckmans, C. Vandecasteele and J. Hoste, *Anal. Chim. Acta*, 1977, **89**, 255.

[20] A. O. Sanni, N. G. Roché, H. J. Dowell, E. A. Schweikert and T. H. Ramsey, *J. Radioanal. Nucl. Chem.*, 1984, **812**, 125.

[21] M. Fedoroff, C. Loos-Neskovic, J. C. Rouchaud and G. Revel, in *Analysis of Non-metals in Metals*, G. Kraft (ed.), p. 243. De Gruyter, Berlin, 1981.

[22] M. Valladon and J. L. Debrun, *J. Radioanal. Chem.*, 1977, **39**, 385.

[23] G. Blondiaux, C. S. Sastri, M. Valladon and J. L. Debrun, *J. Radioanal. Chem.*, 1980, **56**, 163.

[24] G. Blondiaux, A. Giovagnoli, K. Ishii, C. Koemmerer and M. Valladon, *J. Radioanal. Chem.*, 1980, **55**, 407.

[25] G. Blondiaux, *Ph. D. Thesis*, Université d'Orléans, 1980.

[26] M. Valladon, G. Blondiaux, A. Giovagnoli, C. Koemmerer and J. L. Debrun, *Anal. Chim. Acta*, 1980, **116**, 25.

[27] C. Vandecasteele, F. Alluyn, J. Dewaele and R. Dams, *Anal. Chem.*, 1985, **57**, 2549.

[28] C. Vandecasteele, Unpublished results.

[29] Knolls Atomic Power Laboratory, *Chart of the Nuclides*, 1972.

4

Determination of light elements

A large part of the literature on CPAA deals with the determination of the light elements boron, carbon, nitrogen and oxygen. These elements, sometimes even in trace concentrations, have an important influence on the physical and technological properties of metals and semiconductor materials, and there is a lack of methods that allow their sensitive and accurate determination [1]. In industry, classical methods such as combustion in an oxygen stream followed by measurement of carbon dioxide for carbon, the Kjeldahl method or reducing fusion for nitrogen, and reducing fusion for oxygen are commonly used. These methods have detection limits of, at best, $1 \mu g/g$ and systematic errors may occur owing to surface contamination, difficulties with the determination of the blank, and incomplete extraction of the gases.

CPAA is, in general, not subject to these limitations, since blank problems do not exist, and, in the case of massive samples, the sensitivity is not limited by surface contamination, since after the irradiation a surface layer containing the activity formed at the surface can be removed, for instance by chemical etching. The determination of light elements in metals and semiconductors is therefore one of the main areas of application of CPAA.

Some attention has also been given to the determination by CPAA of lithium and boron in materials other than metals or semiconductor materials, for instance rocks.

Most nuclear reactions of light elements lead to pure positron emitters, e.g. ^{15}O, ^{13}N, ^{11}C and ^{18}F. Detection of β^+-emitters by means of annihilation radiation (Section 3.2) is not specific. Therefore, to obtain low detection limits for the elements of interest, it is often necessary to apply a radiochemical separation. To separate ^{15}O from a metal matrix, reducing fusion [2], whereby the sample is fused in a graphite crucible and the oxygen present yields carbon monoxide, is the method of choice. The short half-life of ^{15}O (2.05 min) is, however, an additional complication. For the separation of ^{13}N, distillation of $^{13}NH_3$ (Kjeldahl method) [3,4] or reducing fusion, whereby nitrogen is liberated as nitrogen gas, can be applied. ^{11}C can be isolated as carbon dioxide by combustion of the sample in an oxygen stream [5,6] or by dissolving the sample in an oxidizing acid mixture [7]. To separate ^{18}F from an irradiated sample, steam distillation of fluorosilicic acid from perchloric, sulphuric, or phosphoric acid in the presence of glass or silica is often used. In general, a fluoride carrier is added to the sample before the distillation. The ^{18}F may be detected directly in the distillate or after precipitation of lead chlorofluoride [8], calcium fluoride [3], or lanthanum fluoride. Another way of separating ^{18}F is by selective fixation on columns filled with compounds such as antimony pentoxide [9], alumina [10], hafnium dioxide [11], tin dioxide [12], or hydroxyapatite [13]. Liquid–liquid extraction, for instance with diphenyldichlorosilane [14,15] or triphenylantimony dichloride [16] may also be applied. Nozaki [17] has separated ^{18}F by precipitation as KBF_4, the KBF_4 being purified by recrystallization.

4.1 METALS AND SEMICONDUCTOR MATERIALS

4.1.1 Boron

Table 4.1 gives nuclear reactions that can be used for the determination of boron,

Table 4.1 — Nuclear reactions for the determination of boron. The threshold energy in MeV is given in brackets. (Reproduced by permission, from Ch. Engelmann, *J. Radioanal. Chem.*, 1971, 7, 89, 281. Copyright 1971, Akadémiai Kiadó, Budapest.)

Nuclear reaction		Interference reaction		Sensitivity*	
$^{11}B(p,n)^{11}C$	(3.0)	$^{154}N(p,\alpha)^{11}C$	(3.1)	0.92	(5 MeV)
				0.24	(10)
$^{10}B(d,n)^{11}C$	$(Q>0)$	$^{12}C(d,t)^{11}C$	(14.5)	3.5	(5)
$^{11}B(d,2n)^{11}C$	(5.9)	$^{14}N(d,\alpha n)^{11}C$	(5.8)	2.1	(10)
				0.75	(15)
$^{10}B(^3He,d)^{11}C$	$(Q>0)$	$^9Be(^3He,n)^{11}C$	$(Q>0)$	44	(5)
$^{11}B(^3He,t)^{11}C$	(2.5)	$^{12}C(^3He,\alpha)^{11}C$	$(Q>0)$	8.2	(10)
		$^{14}N(^3He,\alpha d)^{11}C$	(10.2)	3	(15)
		$^{16}O(^3He,2\alpha)^{11}C$	(6.3)	1.3	(20)
$^{10}B(\alpha,n)^{13}N$	$(Q>0)$	$^{12}C(\alpha,t)^{13}N$	(23.8)	20	(26)
$^{11}B(\alpha,2n)^{13}N$	(14.2)	$^{14}N(\alpha,\alpha n)^{13}N$	(13.6)	10	(30)
		$^{16}O(\alpha,\alpha t)^{13}N$	(26.7)	3.5	(38)
		$^{16}O(\alpha,^7Li)^{13}N$	(28.2)		
		$^{19}F(\alpha,2\alpha 2n)^{13}N$	(34.5)		
		$^{19}F(\alpha,^{10}Be)^{13}N$	(19.5)		
$^{10}B(p,\alpha)^7Be$	$(Q>0)$	$^7Li(p,n)^7Be$	(1.8)		
$^{11}B(p,\alpha n)^7Be$	(11.2)	$^9Be(p,t)^7Be$	(14.9)		
		$^{12}C(p,p\alpha n)^7Be$	(28.5)		
		$^{14}N(p,2\alpha)^7Be$	(11.3)		
$^{10}B(d,\alpha n)^7Be$	(1.3)	$^6Li(d,n)^7Be$	$(Q>0)$		
$^{11}B(d,\alpha 2n)^7Be$	(14.8)	$^7Li(d,2n)^7Be$	(5.0)		
		$^9Be(d,tn)^7Be$	(21.5)		
		$^{12}C(d,\alpha p2n)^7Be$	(33.3)		
		$^{14}N(d,2\alpha n)^7Be$	(14.5)		

*Boron concentration in aluminium (ng/g) that yields an activity of 100 dpm at the end of the irradiation (intensity = 1 μA, irradiation time = 1 half-life) at the energy given in brackets [18].

and also lists the nuclear interferences and the sensitivity. The radionuclides produced, ^{13}N, ^{11}C and ^{7}Be, have half-lives of 9.97 min, 20.4 min and 53.28 d, respectively. ^{13}N and ^{11}C are pure positron emitters, ^{7}Be emits 477.6 keV γ-rays.

For the ^{40}B(d,n)^{11}C and the ^{10}B(α,n)^{13}N reactions an energy below the threshold of the interfering reactions (5.8 and 13.6 MeV, respectively) can be used, so that no nuclear interferences occur.

The ^{11}B(p,n)^{11}C reaction yields the highest sensitivity, but interference of nitrogen by the ^{14}N(p,α)^{11}C reaction cannot be avoided.

Figure 4.1 gives the ratio of the ^{11}C activity produced from boron to that

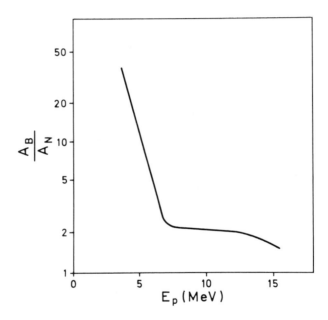

Fig. 4.1 — Ratio of the ^{11}C activity produced from boron to that from the same amount of nitrogen, as a function of the proton energy. (Reproduced by permission, from Ch. Engelmann, *J. Radioanal. Chem.*, 1971, **7**, 281. Copyright 1971, Akadémiai Kiadó, Budapest.)

produced from nitrogen by the ^{14}N(p,α)^{11}C reaction, as a function of the proton energy [19]. The ^{11}C activity from boron is about twice that from nitrogen in the 7–15 MeV energy range, and about ten times at 5 MeV. This corresponds to an interference factor of 0.5 in the 7–15 MeV range and of 0.1 at 5 MeV. When the boron and nitrogen concentrations are comparable, it is in principle possible to determine boron and nitrogen by irradiation of two different samples with protons of different energies, as suggested by Kuin [20] and by Rommel [21]. This method is based on the different energy-dependence of the thick target yields for the ^{11}B(p,n)^{11}C and ^{14}N(p,α)^{11}C reactions and leads to two equations with two unknowns that can be solved for the nitrogen and boron concentrations. Mortier *et al.* [22] described a similar approach for the simultaneous determination of boron and lithium, making use of nuclear reactions leading to ^{7}Be (see Section 4.2.1).

For the ^{10}B(^{3}He,d)^{11}C and ^{10}B(α,n)^{13}N reactions, nuclear interferences cannot be avoided and the sensitivity is rather low.

The ^{10}B(p,α)^{7}Be and ^{10}B(d,αn)^{7}Be reactions have the following advantages: the long half-life of ^{7}Be allows complex chemical separations, and, since ^{7}Be is a γ-ray emitter, instrumental analysis is sometimes possible. Disadvantages are that the interference from lithium cannot be avoided and long irradiation and counting times are required.

The most suitable standard for boron is czochralski boron: single crystals of boron, available in rods of 5–6 mm diameter, from which discs of 1–2 mm thickness can be cut with a diamond saw. When the ^{10}B(d,n)^{11}C reaction is used, this material is an almost ideal standard, because of the high purity ($>99.9\%$), stability under irradiation, and pure ^{11}C-decay. Additional advantages are that the material can be used again after a short time, and the stopping powers are more accurately known for a pure element than for a compound.

Mortier [23] compared several materials that can be used as standards for the determination of boron, using nuclear reactions leading to ^{7}Be. The same results, within experimental error, were obtained for czochralski boron, boron powder, boric acid (H_3BO_3), boron trioxide, and sodium tetraborate ($Na_2B_4O_7$). Boron nitride (BN) yielded somewhat higher, and borax ($Na_2B_4O_7.10H_2O$) significantly higher results. For borax this was due to loss of water during the irradiation. Since the czochralski boron standards could, because of the long half-life of ^{7}Be, only be used again after about 1 year, boric acid was used as standard. Boric acid powder can easily be pelleted, but a disadvantage is that loss of water starts to occur at 80°C. It was experimentally shown that for 12 MeV protons, beam intensities up to 300 nA can be used.

Table 4.2 gives a literature survey of the determination of boron in metals and semiconductor materials. It is clear that in almost all recent publications use is made of the ^{10}B(d,n)^{11}C reaction or of nuclear reactions leading to ^{7}Be.

As an example, the determination of boron in aluminium and aluminium–magnesium alloy by using the ^{10}B(d,n)^{11}C reaction [7] will be described in some detail.

4.1.1.1 Determination of boron in aluminium and aluminium–magnesium alloy by using the ^{10}B(d,n)^{11}C reaction [7]

Aluminium and aluminium–magnesium alloy are used as construction materials in the nuclear industry. Because of the high cross-section of ^{10}B for the capture of thermal neutrons, the boron concentration must be very low. Boron also precipitates titanium from molten aluminium alloys, thus improving the electrical conductivity, an important property for aluminium alloys used in the electrical industry. Titanium boride acts as a seed for the aluminium crystals and thus has a grain-refining action resulting in improved mechanical properties and simplifying casting. Accurate knowledge of the boron concentration is therefore necessary to control the refining process. In order to control the accuracy of the routine methods used in industry to determine boron in aluminium and aluminium–magnesium alloy, certified reference materials are required. The method described by Mortier *et al.* [7] was developed mainly for the analysis of reference materials to be certified by BCR, the Community

Table 4.2 — Literature survey of the determination of boron in metals and semiconductor materials.

Reference		Nuclear reaction	Matrix analysed	Concentration, μg/g
Engelmann *et al.*	[24]	$^{11}B(p,n)^{11}C$	Si	5×10^{-4}–0.2
Shibata *et al.*	[25]	$^{10}B(p,\alpha)^7Be$	Al	2–5
Goethals *et al.*	[26]	$^{10}B(p,\alpha)^7Be$	Al	1.1
		$^{10}B(\alpha,n)^{13}N$	AlMg	66
		$^{10}B(d,n)^{11}C$		
Mortier *et al.*	[27]	$^{10}B(p,\alpha)^7Be$	Zr	<0.015–100
			Zircaloy	0.16
Petit *et al.*	[5]	$^{10}B(d,n)^{11}C$	Zr	24×10^{-3}
Strijckmans *et al.*	[3]	$^{10}B(d,n)^{11}C$	Ni	0.13
Mortier *et al.*	[28]	$^{10}B(d,n)^{11}C$	Zircaloy	0.16
	[29]	$^{10}B(p,\alpha)^7Be$	Zr	<0.03–100
		$^{10}B(d,n)^{11}C$	Zircaloy	0.16
Sastri *et al.*	[30]	$^{10}B(p,\alpha)^7Be$	Nb	0.04–0.2
		$^{10}B(d,\alpha n)^7Be$		
Mortier *et al.*	[7]	$^{10}B(p,\alpha)^7Be$	Al	1.1
		$^{10}B(d,\alpha n)^7Be$	AlMg	33
		$^{10}B(d,n)^{11}C$		

Bureau of Reference of the Commission of the European Communities.

The samples were primary ingot aluminium (BCR CRM No 25; 99.5% purity) and aluminium–magnesium alloy (BCR CRM No 330; 3.2% Mg) in the form of cylindrical discs 15 mm in diameter and 1 mm thick. Pure boron was used as a standard: czochralski boron, available as a single-crystal rod with 5–6 mm diameter, was fixed with 'Araldite' in an aluminium tube and sawn into discs 1 mm thick.

The samples and the standards were irradiated in vacuum, behind a nickel foil serving as a beam intensity monitor. Between the monitor foil and the sample a 4.8 mg/cm^2 aluminium foil was placed to stop recoil nuclei. Such foils were also placed before the standards. Table 4.3 summarizes the irradiation conditions and gives

Table 4.3 — Irradiation conditions

	Sample	Standard
Energy, MeV	7	7
Intensity, μA	2	0.05
Energy after chemical etching, MeV	5.3–5.6	
Irradiation time, min	20	1
Beam intensity monitoring		
Monitor foil	Ni, 11 mg/cm^2	
Nuclear reaction	$^{60}Ni(d,n)^{61}Cu$	
Induced activity, $t_{1/2}$	3.41 hr	
E_γ	283 keV	
Additional foils (4.8 mg/cm^2)	1	1 or 2
Thickness removed by etching (mg/cm^2)	5–8	
Effective incident energy (MeV)	4.7–4.9	
Residual range (mg/cm^2)	29–32	

information on beam intensity monitoring. Before the monitor foil, a diaphragm was placed, with a diameter smaller than that of the sample or the standard.

After the irradiation, the samples were etched for 1 min in a mixture of $15M$ H_3PO_4, $18M$ H_2SO_4, and $14M$ HNO_3 (7:2:1, v/v/v) at 80–90°C, in order to remove a 1.3–2.7 mg/cm^2 surface layer. After etching, the samples were rinsed with water and acetone and dried in a stream of hot air.

Instrumental measurements of ^{11}C by means of annihilation radiation are interfered with, for example, by ^{22}Na and ^{24}Na formed from the matrix. Therefore, ^{11}C is isolated as carbon dioxide by dissolving the sample in an oxidizing acid solution. The set-up is shown in Fig. 4.2. The absorption vessel containing $18M$

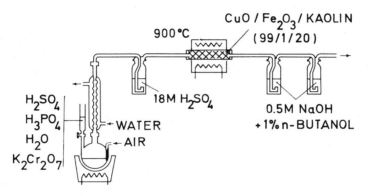

Fig. 4.2 — Set-up for the isolation of ^{11}C as carbon dioxide. (Reprinted with permission, from R. Mortier, C. Vandecasteele, K. Strijckmans and J. Hoste, *Anal. Chem.*, 1984, **56**, 2166. Copyright 1984, American Chemical Society).

H_2SO_4 retains volatile matrix activity. The liberated gases are led over an oxidizing mixture to oxidize any carbon monoxide and hydrocarbons to carbon dioxide. The carbon dioxide is finally trapped in $0.5M$ NaOH. The detailed experimental procedure is as follows. To prepare the oxide mixture, mix kaolin, copper(II) oxide, and iron(III) oxide (20/99/1 by weight) by shaking for 24 hr. Add a minimum amount of water to obtain a paste. Squeeze the paste through a 1 mm orifice. Dry the resulting threads at 100°C, cut them into 3–4 mm lengths, and heat at 900°C for 3–4 hr. Reduce with hydrogen at 250°C for 3 hr and oxidize with oxygen at 400–500°C. Fill a 30-cm long silica tube (15 mm in diameter). To prepare the oxidizing acid solution, heat a mixture of 300 ml of $18M$ H_2SO_4, 200 ml of $15M$ H_3PO_4, and 50 ml of water to 150°C. Gradually add 75 g of $K_2Cr_2O_7$ with stirring during the heating. Place the sample, with 100 mg of graphite carrier, in the reaction vessel. Add 60 ml of the acid solution, pass air through the mixture, and heat to 150–170°C. Lead the gases via the condenser to the scrubbing vessel filled with $18M$ H_2SO_4 and through the tube with the oxide mixture at 900°C to the absorption vessels filled with 100 ml of $0.5M$ NaOH and 1 ml of 1-butanol. After 30 min, ^{11}C is quantitatively absorbed in the first absorption vessel.

To determine the yield of the chemical separation, aluminium doped with 0.5%

boron was irradiated with 4 MeV deuterons. At this concentration level spectral interference from other β^+-emitters was negligible. After the instrumental measurement of the ^{11}C-activity, the chemical separation was carried out. From the activity before and after the separation, a yield of $100.4 \pm 1.8\%$ (mean and standard deviation of 4 experiments) was obtained.

The annihilation radiation of ^{11}C was measured with a γ-γ coincidence set-up consisting of two 7.6×7.6 cm NaI(Tl) detectors in 180° geometry. Because the detection efficiency is different for the standard (a disc source) and the sample (100 ml of solution), a correction factor was determined as described in Section 3.2.1. The radiochemical purity of the sodium hydroxide solution was checked in two ways. A Ge(Li) γ-spectrum showed that no interfering β^+ or γ-emitters were present. In addition, by repetitive measurement with the γ-γ coincidence set-up, a decay curve was obtained. Decay curve analysis indicated pure ^{11}C: the best fit for the half-life was 20.44 ± 0.33 min (mean $\pm$ standard deviation of 8 decay curves for CO_2 from aluminium–magnesium alloy), in agreement with the literature value of 20.4 min.

Table 4.4 gives the results and the certified values. Figures 4.3 and 4.4 compare

Table 4.4 — Determination of boron in aluminium and aluminium–magnesium alloy by using the ^{10}B(d,n)^{11}C reaction. Results in μg/g.

	$\bar{x}$	s	n	Certified value
Aluminium	1.18	0.09	8	1.22 ± 0.06
Aluminium–magnesium alloy	34.7	1.7	8	32 ± 4

the results obtained by the method described with those of other analytical methods applied in the BCR certification campaign leading to the certification of these reference materials. Besides CPAA based on the ^{10}B(d,n)^{11}C reaction, as described, CPAA using the ^{10}B(p,α)^{7}Be and ^{10}B(d,αn)^{7}Be reactions, spectrophotometry by the Methylene Blue method (SPM), inductively-coupled plasma atomic-emission spectrometry (ICPS), isotope-dilution mass-spectrometry (IDMS) and spark-source mass-spectrometry (SSMS) were used. The overall agreement was satisfactory, so certification of the reference materials was possible. For these applications, ICPS and IDMS are more precise than CPAA, probably owing to the different sample masses. ICPS and IDMS analyse 0.5–1 g of sample, CPAA only about 40 mg. The influence of a boron distribution that is not fully homogeneous is therefore more significant in CPAA.

4.1.2 Carbon

Table 4.5 gives nuclear reactions that can be used for carbon determination, along with the nuclear interferences and the sensitivity. The radionuclides produced, ^{13}N and ^{11}C, are positron emitters with half-lives of 9.97 and 20.4 min, respectively. For the ^{12}C(^{3}He,α)^{11}C reaction, nuclear interferences occur that cannot be avoided by appropriate choice of the incident energy. The same holds for the ^{12}C(α,αn)^{11}C reaction. For the ^{12}C(p,γ)^{13}N and ^{13}C(p,n)^{13}N reactions and for the ^{12}C(d,n)^{13}N

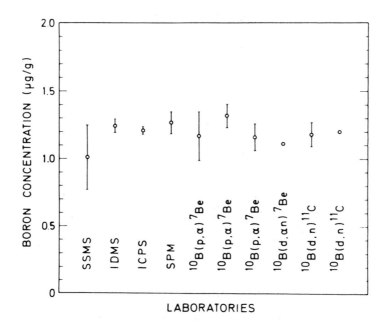

Fig. 4.3 — Comparison of the results for boron in aluminium (mean ± standard deviation), obtained by different methods (see text). (Reprinted with permission, from R. Mortier, C. Vandecasteele, K. Strijckmans and J. Hoste, *Anal. Chem.*, 1984, **56**, 2166. Copyright 1984, American Chemical Society).

reaction, nuclear interferences can be avoided by choosing an incident energy below the threshold of the interfering reactions, 5.5 and 4.9 MeV, respectively. It appears from Table 4.5 that under these conditions the $^{12}C(d,n)^{13}N$ reaction has the highest sensitivity. The excitation function of this reaction is given in Fig. 2.7 (p. 34).

As a standard for carbon, discs of graphite or polyethylene are often used.

Table 4.6 gives a literature survey of the determination of carbon in metals and semiconductor materials. The $^{12}C(d,n)^{13}N$ and $^{12}C(^{3}He,\alpha)^{11}C$ reactions are clearly the most useful nuclear reactions and the most often used. As an example, the determination of carbon in molybdenum and tungsten [6] by the $^{12}C(d,n)^{13}N$ reaction will be described in some detail.

4.1.2.1 Determination of carbon in molybdenum and tungsten by using the $^{12}C(d,n)^{13}N$ reaction [6]

Table 4.7 summarizes the irradiation conditions and gives some information on the chemical etch after the irradiation for the determination of carbon in molybdenum and tungsten by the $^{12}C(d,n)^{13}N$ reaction [6]. After a 15 min delay the samples were repeatedly measured for 3 min during 30–50 min, by means of a γ-γ coincidence set-up.

The incident energy is not critical for tungsten, but very critical for molybdenum. Figure 4.5 gives the activation curves for the $^{92}Mo(d,n)^{93m}Tc$ and $^{100}Mo(d,p)^{101}Mo$

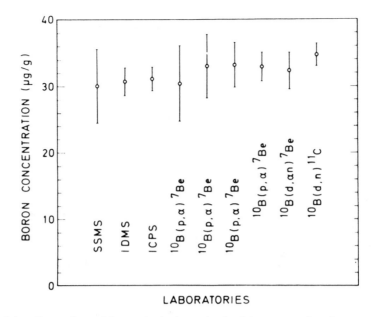

Fig. 4.4 — Comparison of the results for boron in aluminium–magnesium (mean ± standard deviation), obtained by different methods (see text). (Reprinted with permission, from R. Mortier, C. Vandecasteele, K. Strijckmans and J. Hoste, *Anal. Chem.*, 1984, **56**, 2166. Copyright 1984, American Chemical Society).

Table 4.5 — Nuclear reactions for the determination of carbon. The threshold energy in MeV is given in brackets. (Reproduced by permission, from Ch. Engelmann, *J. Radioanal. Chem.*, 1971, 7, 89, 281. Copyright 1971, Akadémiai Kiadó, Budapest.)

Nuclear reaction		Interference reaction		Sensitivity*	
$^{12}C(p,\gamma)^{13}N$	$(Q>0)$	$^{16}O(p,\alpha)^{13}N$	(5.5)	110	(5 MeV)
$^{13}C(p,n)^{13}N$	(3.2)	$^{14}N(p,d)^{13}N$	(8.9)	40	(7.5)
		$^{14}N(p,pn)^{13}N$	(11.3)	18	(15)
$^{12}C(d,n)^{13}N$	(0.3)	$^{14}N(d,t)^{13}N$	(4.9)	7	(5)
		$^{14}N(d,dn)^{13}N$	(12.1)	0.7	(10)
		$^{16}O(d,\alpha n)^{13}N$	(8.4)	0.4	(20)
$^{12}C(^{3}He,\alpha)^{11}C$	$(Q>0)$	$^{9}Be(^{3}He,n)^{11}C$	$(Q>0)$	12	(5)
		$^{10}B(^{3}He,d)^{11}C$	$(Q>0)$	3.6	(10)
		$^{11}B(^{3}He,t)^{11}C$	(2.5)	0.85	(20)
		$^{14}N(^{3}He,\alpha d)^{11}C$	(10.2)		
		$^{16}O(^{3}He,2\alpha)^{11}C$	(6.3)		
$^{12}C(\alpha,\alpha n)^{11}C$	(25)	$^{9}Be(\alpha,2n)^{11}C$	(18.8)	16	(34)
		$^{10}B(\alpha,t)^{11}C$	(15.6)	4.4	(38)
		$^{11}B(\alpha,tn)^{11}C$	(30.8)	1.75	(42)
		$^{14}N(\alpha,\alpha t)^{11}C$	(29.2)		
		$^{16}O(\alpha,2\alpha n)^{11}C$	(32.5)		

*Carbon concentration (ng/g) in an aluminium sample that yields an activity of 100 dpm at the end of the irradiation (intensity = 1 μA, irradiation time = 1 half-life) at the energy given in brackets [18].

Table 4.6 — Literature survey of the determination of carbon in metals and semiconductor materials.

Reference		Nuclear reaction	Matrix analysed	Concentration, μg/g
Albert et al.	[31]	$^{12}C(d,n)^{13}N$	Fe	2–90
Albert et al.	[32]	$^{12}C(d,n)^{13}N$	Al	
Nozaki et al.	[33]	$^{12}C(^{3}He,\alpha)^{11}C$	Si	20×10^{-3}–2
Engelmann and Marschal	[34]	$^{12}C(d,n)^{13}N$ $^{12}C(^{3}He,\alpha)^{11}C$ $^{12}C(\alpha,\alpha n)^{11}C$	Si	0.2–0.3
Endo et al.	[35]	$^{12}C(^{3}He,\alpha)^{11}C$	Si	60×10^{-3}–3
Mayolet et al.	[36]	$^{12}C(\alpha,n)^{15}O$	Ta	8
			Fe	2
Martin and Haas	[37]	$^{12}C(d,n)^{13}N$	Si	0.2–130
Nozaki et al.	[38]	$^{12}C(^{3}He,\alpha)^{11}C$	Si	10×10^{-3}
Vandecasteele et al.	[39]	$^{12}C(^{3}He,\alpha)^{11}C$	Si	0.1–3
Nozaki et al.	[40]	$^{12}C(^{3}He,\alpha)^{11}C$	Si	
Goethals et al.	[4]	$^{12}C(d,n)^{13}N$	Al	0.3
Vandecasteele et al.	[41]	$^{12}C(d,n)^{13}N$	Zr	65
			Nb	25
			Ta	1
			W	$< 15 \times 10^{-3}$
Valladon et al.	[42]	$^{12}C(d,n)^{13}N$	GaAs	22×10^{-3}–0.4
Strijckmans et al.	[3]	$^{12}C(d,n)^{13}N$	Ni	85
Vandecasteele et al.	[6]	$^{12}C(d,n)^{13}N$	Mo	70×10^{-3}–0.1
			W	8×10^{-3}
Mortier et al.	[28]	$^{12}C(d,n)^{13}N$	Zircaloy	110
Sanni et al.	[43]	$^{12}C(d,n)^{13}N$	Si	0.005–30
Vandecasteele et al.	[44]	$^{12}C(d,n)^{13}N$	Au	2–1300
Böttger et al.	[45]	$^{12}C(d,n)^{13}N$ $^{12}C(^{3}He,\alpha)^{11}C$	Si	10×10^{-3}–20×10^{-3}

Table 4.7 — Irradiation and post-irradiation chemical etch.

	Molybdenum	Tungsten
Deuteron energy (MeV)	5	5
Intensity (μA)	2	2
Irradiation time (min)	20	20
Monitor foil	Mo 9.34 mg/cm^2	Mo 9.34 mg/cm^2
Etch	4:1 (v/v) 40% HF/14M HNO$_3$ 70 sec, room temp.	1:1 (v/v) 50% HF/14M HNO$_3$ 100 sec, room temp.
Thickness removed (mg/cm^2)	23–28	34–40
Effective incident energy (MeV)	2.3–2.7	2.3–2.7
Residual range (mg/cm^2)	16–21	28–34

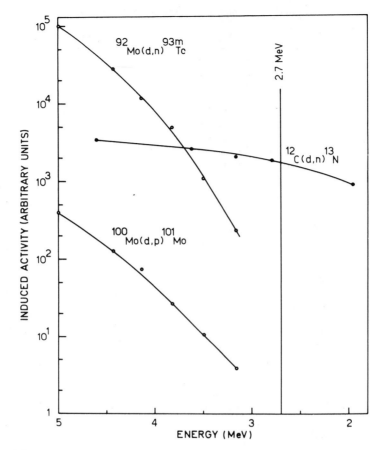

Fig. 4.5 — Activation curve for the ^{92}Mo(d,n)^{93m}Tc ($E_\gamma = 390.0$ keV, $t_{1/2} = 43$ min), ^{100}Mo(d,p)^{101}Mo ($E_\gamma = 1012.4$ keV, $t_{1/2} = 14.6$ min), and ^{12}C(d,n)^{13}N reactions. (Reproduced by permission, from C. Vandecasteele, K. Strijckmans, Ch. Engelmann and H. M. Ortner, *Talanta*, 1981, **28**, 19. Copyright 1981, Pergamon Press, Oxford).

reactions, determined experimentally by irradiation of a stack of molybdenum foils with 5 MeV deuterons. For comparison, the activation curve of the ^{12}C(d,n)^{13}N reaction is also given. The activity produced from molybdenum decreases rapidly with decreasing energy: below 3 MeV, the matrix activity is less than 0.2% of that at 5 MeV. For an incident energy between 2.3 and 2.7 MeV an instrumental analysis is feasible.

For molybdenum the decay curve consists of two components: ^{13}N and ^{93}Tc ($t_{1/2} = 2.73$ hr) formed by the ^{92}Mo(d,n)^{93}Tc reaction; for tungsten it is a one-component curve with only ^{13}N. The decay curves were analysed by the CLSQ decay-curve analysis program, the best fitting half-life for the short-lived component being on the average 8.5 min with a standard deviation of 0.8 min (molybdenum) and 10.7 min with a standard deviation of 1.8 min (tungsten).

Table 4.8 gives the results and compares them with those obtained by photon

Table 4.8 — Determination of carbon in molybdenum and tungsten. Results in $\mu g/g$.

	Deuteron activation		Photon activation		Combustion
	$\bar{x}$	s	$\bar{x}$	s	$\bar{x}$
Molybdenum					
BCR	0.098	0.018	0.063	0.050	
Plansee	0.097	0.044			
GDMB	0.077	0.021	0.13	0.06	5–18
Tungsten					
Plansee	0.0084	0.0024			
GDMB	0.0114	0.0041	0.03	0.03	3–9

activation and by the combustion method. In the period 1969–1971 the 'Gesellschaft Deutscher Metallhütten und Bergleute' (GDMB) organized a first interlaboratory round-robin relative to the determination of carbon in massive samples of molybdenum. All participating laboratories applied different variants of the combustion method. The mean values [46] ranged from 1 to 10 $\mu g/g$, so it was concluded that the combustion method is not suited to determine carbon concentrations below 10 $\mu g/g$ in molybdenum. After an additional study of the possible sources of error, in 1976 a second round-robin yielded results between 5 and 18 $\mu g/g$ for molybdenum and between 3 and 9 $\mu g/g$ for tungsten. It was concluded [46] that the carbon concentration was ca. 15 $\mu g/g$ for molybdenum and ca. 5 $\mu g/g$ for tungsten. Differences between laboratories were attributed to inhomogeneity of the material analysed and to the different experimental conditions: high-frequency induction furnace or resistance furnace, combustion temperature (1250–2000°C), sample weight, nature and quantity of fluxes added, calibration. It appears, however, from Table 4.8 that the results obtained by deuteron activation analysis are more than 2 orders of magnitude lower than those obtained by the combustion method, but are in agreement with those obtained by photon activation analysis [6,47]. Deuteron activation yields, however, at this concentration level, more precise results than photon activation, and has the advantage of being purely instrumental, whereas photon activation requires a chemical separation of the ^{11}C produced by the $^{12}C(\gamma,n)^{11}C$ reaction. From a technical point of view these concentrations are surprisingly low. It was not thought previously that molybdenum and tungsten with such low concentrations of carbon could be produced by powder metallurgy on an industrial scale, without special precautions. The samples analysed were indeed taken from normal production of sheet material. A possible explanation for the high results obtained by the combustion method is corrosion of the crucible material by the acidic molybdenum and tungsten oxide melts, whereby carbon dioxide was liberated [6].

4.1.3 Nitrogen
Table 4.9 gives nuclear reactions that can be used for nitrogen determination, along with the nuclear interferences and the sensitivity. The radionuclides produced are positron emitters with a half-life of 66.0 sec (^{17}F), 70.5 sec (^{14}O), 2.03 min (^{15}O), 9.97 min (^{13}N), 20.4 min (^{11}C) and 109.8 min (^{18}F). ^{14}O also emits γ-rays of 2313.0 keV.

Table 4.9 — Nuclear reactions for the determination of nitrogen. The threshold energy in MeV is given in brackets. (Reproduced by permission, from Ch. Engelmann, *J. Radioanal. Chem.*, 1971, **7**, 89, 281. Copyright 1971, Akadémiai Kiadó, Budapest.)

Nuclear reaction		Interference reaction		Sensitivity*	
$^{14}N(p,n)^{14}O$	(6.3)	$^{16}O(p,t)^{14}O$	(21.7)		
$^{14}N(p,\alpha)^{11}C$	(3.1)	$^{11}B(p,n)^{11}C$	(3.0)	0.50	(10 MeV)
		$^{12}C(p,d)^{11}C$	(17.9)	0.22	(15)
$^{14}N(d,n)^{15}O$	($Q>0$)	$^{16}O(d,t)^{15}O$	(10.6)	0.8	(10)
				0.5	(15)
$^{14}N(d,\alpha n)^{11}C$	(5.8)	$^{10}B(d,n)^{11}C$	($Q>0$)	2.7	(15)
		$^{12}C(d,t)^{11}C$	(14.5)	0.75	(20)
$^{14}N(^{3}He,\alpha)^{13}N$	($Q>0$)	$^{11}B(^{3}He,n)^{13}N$	($Q>0$)	66	(10)
		$^{12}C(^{3}He,d)^{13}N$	(4.5)	29	(20)
$^{14}N(^{3}He,d)^{15}O$	($Q>0$)	$^{13}C(^{3}He,n)^{15}O$	($Q>0$)		
		$^{16}O(^{3}He,\alpha)^{15}O$	($Q>0$)		
$^{14}N(\alpha,n)^{17}F$	(6.0)	$^{16}O(\alpha,t)^{17}F$	(24.0)		
$^{15}N(\alpha,n)^{18}F$	(8.1)	$^{16}O(\alpha,d)^{18}F$	(20.4)		
$^{14}N(\alpha,\alpha n)^{13}N$	(13.6)	$^{10}B(\alpha,n)^{13}N$	($Q>0$)	12	(26)
		$^{12}C(\alpha,t)^{13}N$	(23.9)	3.2	(34)

*Nitrogen concentration (ng/g) in an aluminium sample, that gives an activity of 100 dpm at the end of the irradiation (intensity = 1 μA, irradiation time = 1 half-life) at the energy in brackets [18].

For the $^{14}N(p,n)^{14}O$ reaction, no nuclear interferences occur at a proton energy below 21.7 MeV. In addition, spectral interferences are rather unlikely when the 2313.0 keV γ-rays are measured. A disadvantage is the high threshold energy (6.3 MeV) which makes it, for low Z matrices, impossible to avoid activation of the matrix.

The $^{14}N(p,\alpha)^{11}C$ reaction is interfered with by boron. Since the boron concentration in industrial metals is usually lower than the nitrogen concentration, a correction can be deduced, where necessary, from the boron concentration determined as described in Section 4.1.1. Figures 2.8–2.10 (p. 35,36) give the excitation function for the $^{14}N(p,\alpha)^{11}C$ reaction.

For the $^{14}N(d,n)^{15}O$ reaction, nuclear interferences can be avoided by limiting the energy to below 10.6 MeV. This method has a high sensitivity, but, because of the short half-life, it is difficult to separate ^{15}O without a large decrease in the sensitivity.

The $^{14}N(d,\alpha n)^{11}C$ reaction has a lower sensitivity than the $^{14}N(p,\alpha)^{11}C$ reaction, and the interference from boron is more important [19], so the $^{14}N(p,\alpha)^{11}C$ reaction must be preferred.

As a nitrogen standard, nylon or compounds such as sodium nitrate can be used.

Table 4.10 gives a literature survey of the determination of nitrogen in metals and semiconductors. It is clear that the $^{14}N(p,\alpha)^{11}C$ reaction is most often applied.

As an example, the determination of nitrogen in zirconium and zircaloy [28,55] will be described in some detail.

Table 4.10 — Literature survey of the determination of nitrogen in metals and semiconductor materials.

Reference		Nuclear reaction	Matrix analysed	Concentration, $\mu g/g$
Nozaki et al.	[48]	$^{14}N(p,\alpha)^{11}C$	C	
Nozaki et al.	[33]	$^{14}N(p,\alpha)^{11}C$	Si	20×10^{-3}
Mayolet et al.	[36]	$^{14}N(d,n)^{15}O$	Fe	2–200
			Mo	1–3
Nozaki et al.	[40]	$^{14}N(p,\alpha)^{11}C$	Si	
Strijckmans et al.	[49]	$^{14}N(p,n)^{14}O$	Ta	9
			Nb	200
			Ti	110
Strijckmans et al.	[50]	$^{14}N(p,\alpha)^{11}C$	Ta	11
Petit et al.	[51]	$^{14}N(p,\alpha)^{11}C$	Zr	34
Giovagnoli et al.	[52]	$^{14}N(d,n)^{15}O$	Zr	35
Vandecasteele et al.	[53]	$^{14}N(p,n)^{14}O$	Al–Ti alloy	35–230
Strijckmans et al.	[3]	$^{14}N(p,\alpha)^{11}C$	Ni	1
Vandecasteele et al.	[6]	$^{14}N(p,\alpha)^{11}C$	Mo	0.5
			W	70×10^{-3}
Mortier et al.	[28]	$^{14}N(\alpha,\alpha n)^{13}N$	Zircaloy	37
Sastri et al.	[54]	$^{14}N(p,2\alpha)^{7}Be$	Nb	20–30
		$^{14}N(d,2\alpha n)^{7}Be$		
Strijckmans et al.	[55]	$^{14}N(\alpha,\alpha n)^{13}N$	Zr	10–25
		$^{14}N(p,\alpha)^{11}C$		

4.1.3.1 Determination of nitrogen in zirconium and zircaloy by using the $^{14}N(p,\alpha)^{11}C$ reaction [28,55]

Table 4.11 summarizes the irradiation conditions and the conditions for the chemical etch after the irradiation.

Table 4.11 — Irradiation and post-irradiation chemical etch for the determination of nitrogen in zirconium and zircaloy.

	Sample	Standard (nylon)
Proton energy (MeV)	15	15
Intensity (μA)	1	0.05
Irradiation time (min)	20	1
Monitor foil	Cu 28.5 mg/cm^2	Cu 28.5 mg/cm^2
Additional foils	—	Al 4.7 mg/cm^2
		Al 9.4 mg/cm^2
		Al 14.1 mg/cm^2
Etch	1:40 (v/v) 50% HF/H$_2$O 3 min, room temperature	
Thickness removed (mg/cm^2)	9–16	
Effective incident energy (MeV)	14.1–14.2	14.1–14.3
Residual range (mg/cm^2)	432–437	

   To isolate ^{11}C as carbon dioxide from the matrix the following chemical separation was carried out. Dissolve the sample and about 100 mg of graphite carrier

in a boiling solution of 6 g of sodium fluoride and 6 g of potassium periodate in 60 ml
of concentrated sulphuric acid (Fig. 4.6). Pass air through the solution and lead the

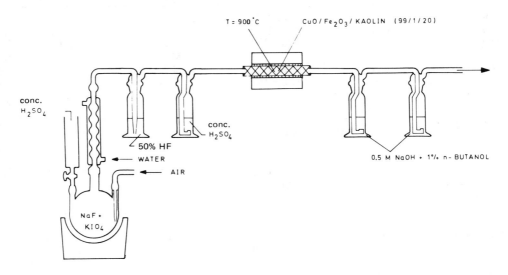

Fig. 4.6 — Chemical separation of $^{11}CO_2$ from zirconium and zircaloy.

liberated gases through a reflux condenser, a plastic absorption vessel containing
50% hydrofluoric acid, an absorption vessel containing concentrated sulphuric acid,
a furnace containing a CuO/Fe_2O_3/kaolin mixture (weight ratio 99/1/20) at 900°C,
and two absorption vessels containing $0.5M$ sodium hydroxide ($+1\%$ v/v
n-butanol). Volatile matrix activity is retained in the hydrofluoric acid trap, the gases
are dried in the sulphuric acid trap, and carbon monoxide and hydrocarbons are
oxidized to carbon dioxide in the furnace. After complete dissolution of the sample
(10 min) pass air through for another 10 min. Carbon dioxide is quantitatively
absorbed in the first absorption vessel with sodium hydroxide.

The contents of the absorption vessels were measured with a γ-γ coincidence set-
up, in order to detect the annihilation radiation from ^{11}C. Measurements started
50–60 min after the irradiation and every 10 min (during the next 60 min) a 2-min
measurement was made in order to obtain the decay curve. For the standards the
measurements started 60–90 min after the irradiation.

A γ-ray spectrum obtained with a semiconductor detector showed only annihilation radiation. The best half-life for ^{11}C was on the average 20.39 min with a standard
deviation of 0.06 min. For the nylon standards the decay curve was composed of
three components: ^{13}N, ^{11}C and ^{18}F.

Because of the low boron concentration the interference from boron was
negligible.

Table 4.12 gives results for three certified reference materials from BCR. In each
case, the agreement with the certified values is satisfactory. During the certification
campaign for nitrogen in zirconium the following methods were applied: reducing
fusion under vacuum and under inert gas, the Kjeldahl method, photon activation,

Table 4.12 — Determination of nitrogen in zirconium and zircaloy. Results in $\mu g/g$.

		Nitrogen concentration		Certified
		$\bar{x}$	s	
Zirconium	BCR 21	25.85	0.79	26.6 ± 2.7
	BCR 56	10.52	0.25	11.7 ± 1.8
Zircaloy	BCR 275	36.3	1.9	38.2 ± 2.8

deuteron activation, α-particle activation and proton activation as described. The last method was one of the most precise methods for each material, as appears from Fig. 4.7, which summarizes the results of the intercomparison [56].

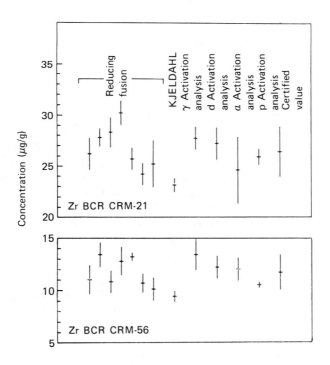

Fig. 4.7 — Intercomparison for the determination of nitrogen in zirconium BCR CRM 21 and 56 (certified value ± uncertainty: 26.6 ± 2.7 $\mu g/g$ and 11.7 ± 1.8 $\mu g/g$).

4.1.4 Oxygen
Table 4.13 gives the most sensitive nuclear reactions for oxygen determination, along with the nuclear interferences. The radionuclides produced, ^{13}N and ^{18}F, are pure positron emitters with a half-life of 9.97 min and 109.8 min, respectively. For the first three groups of nuclear reactions nuclear interferences cannot be avoided by

Table 4.13 — Nuclear reactions for the determination of oxygen. The threshold energy in MeV is given in brackets. (Reproduced by permission, from Ch. Engelmann, *J. Radioanal. Chem.*, 1971, **7**, 89, 281. Copyright 1971, Akadémiai Kiadó, Budapest.)

Nuclear reaction		Interference reaction		Sensitivity* (MeV)	
$^{16}O(p,\alpha)^{13}N$	(5.5)	$^{12}C(p,\gamma)^{13}N$	$(Q>0)$	3	(10)
		$^{13}C(p,n)^{13}N$	(3.2)	1	(15)
		$^{14}N(p,d)^{13}N$	(8.9)		
$^{16}O(^3He,p)^{18}F$	$(Q>0)$	$^{19}F(^3He,\alpha)^{18}F$	$(Q>0)$	2.7	(10)
$^{16}O(^3He,n)$ $^{18}Ne \xrightarrow{\beta+} {}^{18}F$	(3.8)	$^{23}Na(^3He,2\alpha)^{18}F$	(0.4)	1.7	(15)
				1.1	(20)
$^{16}O(\alpha,pn)^{18}F$	(23.2)	$^{15}N(\alpha,n)^{18}F$	(8.1)	2.1	(34)
$^{16}O(\alpha,d)^{18}F$	(20.4)	$^{19}F(\alpha,\alpha n)^{18}F$	(12.6)	1.0	(38)
$^{16}O(\alpha,2n)^{18}Ne \xrightarrow{\beta+} {}^{18}F$	(29.7)	$^{23}Na(\alpha,2\alpha n)^{18}F$	(25)	0.7	(42)
$^{16}O(t,n)^{18}F$	$(Q>0)$	$^{19}F(t,n)^{18}F$	(12.1)		
		$^{24}Mg(t,2\alpha n)^{18}F$	(14.4)		
		$^{20}Ne(t,\alpha n)^{18}F$	(4.0)		

*Oxygen concentration (ng/g) in an aluminium sample that gives an activity of 100 dpm at the end of the irradiation (intensity = 1 μA, irradiation time = 1 half-life) at the energy in brackets [18].

appropriate choice of the energy. Table 4.14 gives the interference factors at different energies [18]. It is clear that helium-3 activation for oxygen is the most selective of the three methods considered. Activation with tritons by the $^{16}O(t,n)^{18}F$ reaction is free from nuclear interferences at below 4.0 MeV. Triton irradiations are, however, only carried out with a limited number of accelerators. The most suitable standard for the determination of oxygen is quartz, which can be obtained with very high purity and is stable under irradiation. When quartz is irradiated with ^{3}He or ^{4}He, a pure ^{18}F-decay is obtained 2–3 hr after the irradiation.

Table 4.14 — Nuclear interferences for the determination of oxygen. (Reproduced by permission, from Ch. Engelmann, *J. Radioanal. Chem.*, 1971, **7**, 89, 281. Copyright 1971, Akadémiai Kiadó, Budapest.)

Method	Energy, MeV	Interference factor			
		C	N	F	Na
Proton activation	10	0.12			
	15	0.050	0.38		
^{3}He-activation	10			0.055	0.020
	15			0.050	0.040
	20			0.058	0.067
α-Activation	35		0.0059	2.50	0.014
	40		0.0040	1.72	0.043

Table 4.15 gives a literature survey of the determination of oxygen in metals and semiconductor materials.

Table 4.15 — Literature survey of the determination of oxygen in metals and semiconductor materials.

Reference		Nuclear reaction	Matrix analysed	Concentration, μg/g
Saito et al.	[57]	$^{16}O(\alpha,pn)^{18}F$	Si	1
Revel and Albert	[58]	$^{16}O(^{3}He,p)^{18}F$	Mo	0.5
		$^{16}O(\alpha,pn)^{18}F$	Zr	3–300
			Hf	8.4
			W	60
Debrun et al.	[59]	$^{16}O(\alpha,pn)^{18}F$	Fe	<0.5–4
		$^{16}O(^{3}He,p)^{18}F$	Ni	<0.5–17
			Cr	<1–115
Schweikert and Rook	[60]	$^{16}O(^{3}He,p)^{18}F$	Si	0.01–15
		$^{16}O(\alpha,pn)^{18}F$		
		$^{18}O(p,n)^{18}F$		
Lee et al.	[61]	$^{16}O(^{3}He,p)^{18}F$	Cu	550
Nozaki et al.	[33]	$^{16}O(^{3}He,p)^{18}F$	Si	5×10^{-3}–10
Engelmann et al.	[24]	$^{16}O(^{3}He,p)^{18}F$	Si	8–12
		$^{16}O(\alpha,pn)^{18}F$		
Engelmann	[62]	$^{18}O(p,n)^{18}F$	Na	—
Kim	[63]	$^{16}O(^{3}He,p)^{18}F$	GaP	5×10^{17}–2×10^{19} at/cm³
			Si	3–36
Faure et al.	[16]	$^{16}O(\alpha,pn)^{18}F$	Pb	0.5
Faure et al.	[15]	$^{16}O(\alpha,pn)^{18}F$	Mo	1–350
Vialatte	[64]	$^{16}O(\alpha,pn)^{18}F$	Al	0.04–0.5
		$^{18}O(p,n)^{18}F$		
		$^{16}O(p,\alpha)^{13}N$		
		$^{16}O(^{3}He,p)^{18}F$		
Vandecasteele et al.	[65]	$^{16}O(\alpha,pn)^{18}F$	Si	4–11
		$^{16}O(^{3}He,p)^{18}F$	Si	0.07–0.3
Fedoroff et al.	[66]	$^{16}O(^{3}He,p)^{18}F$	AlMg	0–12
			Mo	9
			Fe	0.56
Schweikert et al.	[67]	$^{18}O(p,n)^{18}F$	Si	0.006–15
		$^{16}O(^{3}He,p)^{18}F$	Ge	0.03–0.07
		$^{16}O(\alpha,pn)^{18}F$		
Kohn et al.	[11]	$^{16}O(^{3}He,p)^{18}F$	Cu	1–3
		$^{16}O(\alpha,pn)^{18}F$	Co	0.2–700
			Ag	30–400
Vandecasteele and Hoste	[68]	$^{16}O(^{3}He,p)^{18}F$	Ge	0.02–0.7
Vandecasteele et al.	[69]	$^{16}O(\alpha,pn)^{18}F$	Cu	1–300
Valladon and Debrun	[70]	$^{16}O(t,n)^{18}F$	Ti	600
			Mo	16
			Si	8
			Ge	<0.025
			GaAs	<0.006
Vandecasteele and Hoste	[71]	$^{16}O(\alpha,pn)^{18}F$	Pb	0.9
		$^{16}O(^{3}He,p)^{18}F$		
Vandecasteele and Hoste	[8]	$^{16}O(^{3}He,p)^{18}F$	Cu	1
Vandecasteele et al.	[72]	$^{16}O(^{3}He,p)^{18}F$	Al	0.03
			AlSi	0.06
Petri and Sastri	[73]	$^{16}O(^{3}He,p)^{18}F$	Al	
Nozaki et al.	[40]	$^{16}O(^{3}He,p)^{18}F$	Si	
Valladon et al.	[74]	$^{16}O(t,n)^{18}F$	Al	<0.014–0.040
Debefve et al.	[75]	$^{16}O(\alpha,pn)^{18}F$	Au	0.06–1
			Cu	0.2–5
Nozaki et al.	[76]	$^{16}O(^{3}He,p)^{18}F$	SiN film	
Strijckmans et al.	[3]	$^{16}O(^{3}He,p)^{18}F$	Ni	9
Valladon et al.	[42]	$^{16}O(t,n)^{18}F$	GaAs	20×10^{-3}–0.12
Petri and Erdtmann	[77]	$^{16}O(^{3}He,p)^{18}F$	Pt	3
Fedoroff et al.	[78]	$^{16}O(^{3}He,p)^{18}F$	Al	—
Vandecasteele et al.	[6]	$^{16}O(^{3}He,p)^{18}F$	W	0.07
Sanni et al.	[43]	$^{16}O(^{3}He,p)^{18}F$	Si	0.01–15
Debrun and Barrandon	[79]	$^{16}O(^{3}He,p)^{18}F$	Cu	0.3–1.3
			CuZn	2.5
			Ta	2.2
			W	2.3

The determination of oxygen in tungsten by helium-3 activation analysis [6] and of oxygen in aluminium by triton activation analysis [74] will be described in detail.

4.1.4.1 Determination of oxygen in tungsten by helium-3 activation [6]

Table 4.16 gives the irradiation conditions and the conditions for the chemical etch after the irradiation.

To separate ^{18}F from the irradiated sample, the sample is heated with 10 g of sodium hydroxide, 400 mg of sodium fluoride and 2.5 g of sodium nitrate in a nickel crucible, until dissolution is complete. The melt is cooled and transferred into a beaker, and 36 ml of water and 30 ml of $14M$ nitric acid are added in three portions, to dissolve the cooled melt. The solution is transferred into a steam distillation outfit, 25 ml of 85% phosphoric acid are added, the solution is heated to 130°C, steam is introduced, and 200 ml of distillate are collected. The pH is adjusted to 5–6 by addition of $14M$ ammonia and 10 ml of $1M$ sodium carbonate are added. The mixture is heated to 80°C and 20 ml of $2M$ calcium chloride are added with stirring. The calcium fluoride–calcium carbonate precipitate is filtered off and packed in aluminium foil. The sample is repeatedly measured for 30 min with a γ-γ coincidence set-up with two NaI(Tl) detectors, starting 170–200 min after the irradiation. The decay curve found was a one-component curve with a best half-life of 108.8 min and standard deviation of 15.6 min.

Table 4.16 — Irradiation and chemical etch after irradiation for oxygen in tungsten.

	Sample	Standard (quartz)
^{3}He-energy (MeV)	20	20
Intensity (μA)	1.5	0.1
Irradiation time	20 min	20 sec
Monitor foil	Cu 28.4 mg/cm^2	Cu 28.4 mg/cm^2
Additional foils	—	Al 4.9 mg/cm^2
		Al 7.6 mg/cm^2
		Al 9.7 mg/cm^2
Etch	1:1 (v/v) 40% HF/$14M$ HNO$_3$	
	3 min, room temperature	
Thickness removed (mg/cm^2)	12–17	
Effective incident energy (MeV)	13.5–14.2	13.2–14.4
Residual range (mg/cm^2)	74–80	

The yield of the chemical separation is determined by neutron activation with an isotope neutron source and use of the $^{19}F(n,\alpha)^{16}N$ reaction. The precipitate is irradiated with a ^{227}Ac–Be isotope neutron source with a total neutron output of 10^8 neutrons/sec and the induced ^{16}N activity ($t_{1/2} = 7.11$ sec; $E_\gamma = 6.13$ and 7.12 MeV) is measured by means of a cylindrical NaI(Tl) well-type detector 7.6 cm in diameter and 7.6 cm high, the energy range above 4.5 MeV being selected. The irradiation and measurement cycle is fully automatic, controlled by electronic timers: after a 20-sec irradiation, the sample is pneumatically transferred to the measuring position and a 20-sec measurement is started 10 sec after the irradiation. The amount of fluorine is determined by comparison with a sodium fluoride standard.

Table 4.17 compares the results with those obtained by photon activation analysis. In view of the low concentration (the detection limit of reducing fusion for oxygen in tungsten is only ca. 5 μg/g), the agreement is excellent.

Table 4.17 — Results (μg/g) for oxygen in tungsten [6].

^{3}He-activation			Photon activation		
$\bar{x}$	s	n	$\bar{x}$	s	n
0.066	0.011	6	0.074	0.007	7

4.1.4.2 Determination of oxygen in aluminium by triton activation [74]

To determine oxygen in aluminium, Valladon *et al.* [74] used triton activation. The samples were irradiated for 1–2 hr with a 1-μA beam of 3.5 MeV tritons. After the irradiation a 15-μm thick surface layer was removed by a combination of chemical etching and mechanical grinding: first a 2-μm layer was removed by etching in a mixture of hydrofluoric and nitric acids, then 10–12 μm by grinding with silicon carbide paper, and finally another 2 μm by chemical etching. The aluminium sample is very radioactive, because of the ^{27}Al(t,p)^{29}Al and ^{27}Al(t,d)^{28}Al reactions. Since the radionuclides formed have a short half-life (6.52 and 2.24 min respectively), they do not interfere with the detection of ^{18}F after a sufficient waiting time. Neutrons produced by (t,n) reactions with aluminium yield, however, ^{24}Na by the ^{27}Al(n,α)^{24}Na reaction. Because the activity of ^{24}Na is much higher than that of ^{18}F, ^{24}Na must be eliminated before ^{18}F can be detected. For the chemical separation, the sample was dissolved in a 2:1 (v/v) mixture of hydrochloric and nitric acids, in the presence of hydrofluoric acid as carrier. The solution was passed through a column of hydrated antimony pentoxide to trap the sodium. Finally, ^{18}F was measured and the activity compared to that of a pellet of sintered alumina irradiated under similar conditions. The oxygen concentration was below the detection limit, which ranged from 14 to 40 ng/g, depending on the exact experimental conditions.

Activation analysis with ^{3}He also yielded very low results for oxygen in aluminium [72].

In 1969 the Eurisotop Office of the Commission of the European Communities organized a round-robin on the determination of oxygen in industrial aluminium, which was, at that time, considered a very important analytical topic. Initially, reducing fusion and activation analysis with 14 MeV neutrons yielded concentrations between 1 and 20 μg/g. Later on, it was recognized that these methods are subject to systematic errors due to surface contamination and to recoil of ^{16}N nuclei formed by the ^{16}O(n,p)^{16}N reaction from the air around the sample [80,81]. When measures had been taken to overcome these sources of error, it appeared that the detection limits of these methods were too high to allow the determination of oxygen in aluminium. Activation analysis with charged particles [72,74,78] was the first method that showed that the actual oxygen concentration of industrial aluminium is below 0.1 μg/g. This was later on confirmed by other analytical methods. Figure 4.8 shows the oxygen concentration found in a batch of aluminium, as a function of the time at which the analyses were made. The very important decrease in 1976–1977 is mainly due to the careful application of charged particle activation analysis.

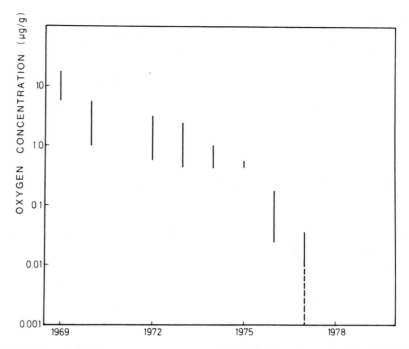

Fig. 4.8 — Trend of the 'apparent oxygen concentration' of primary ingot aluminium [82]. With permission from the author.

4.2   OTHER MATERIALS

There is only a limited interest in the determination of carbon, nitrogen, and oxygen by CPAA in materials other than metals and semiconductors. These elements are usually present in fairly high concentrations in such materials, so simpler analytical techniques can be used for their determination. Some attention has been given to the determination of lithium and boron in rocks [23,83] and in magnesium oxide powder [22].

   Table 4.18 gives nuclear reactions that can be used for the determination of lithium. ^7Be has a half-life of 53.28 d and emits γ-rays of 477.6 keV.

Table **4.18** — Nuclear reactions for the determination of lithium. The threshold energy in MeV is given in brackets.

Nuclear reaction		Interference reactions	
$^7Li(p,n)^7Be$	(1.8)	$^{10}B(p,\alpha)^7Be$	$(Q>0)$
		$^{11}B(p,\alpha n)^7Be$	(11.2)
		$^9Be(p,t)^7Be$	(14.9)
		$^{12}C(p,p\alpha n)^7Be$	(28.5)
		$^{14}N(p,2\alpha)^7Be$	(11.3)
$^6Li(d,n)^7Be$	$(Q>0)$	$^{10}B(d,\alpha n)^7Be$	(1.3)
$^7Li(d,2n)^7Be$	(5.0)	$^{11}B(d,\alpha 2n)^7Be$	(14.8)
		$^9Be(d,tn)^7Be$	(21.5)
		$^{12}C(d,\alpha p2n)^7Be$	(33.3)
		$^{14}N(d,2\alpha n)^7Be$	(14.5)

Vialette [84] proposed the use of the ^{7}Li(p,n)^{7}Be and ^{6}Li(d,n)^{7}Be reactions for the determination of lithium. According to this author, for protons and deuterons with an energy below 5 MeV, the ^{7}Be activity from lithium is lower than the ^{7}Be activity from the same boron concentration. At an energy above 8 MeV the activity from lithium is about 1.4 and about 3 times higher than that from boron, for activation with protons and deuterons, respectively.

Sastri *et al.* [85], however, showed that the ^{7}Be-activity from lithium is about 10 times higher than that from boron, for 4–14 MeV protons. For 4–7 MeV deuterons the activity from lithium is lower than that from boron, but increases more rapidly with the energy. These findings were confirmed by Mortier *et al.* [7] who determined experimentally the interference factor of lithium in the determination of boron, as a function of the proton and deuteron energy (Figs. 4.9 and 4.10).

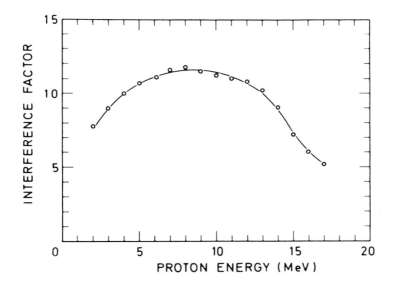

Fig. 4.9 — Interference of lithium with the determination of boron by the ^{10}B(p,α)^{7}Be reaction. (Reprinted with permission, from R. Mortier, C. Vandecasteele, K. Strijckmans and J. Hoste, *Anal. Chem.*, 1984, **56**, 2166. Copyright 1984, American Chemical Society).

For the determination of lithium in the presence of boron the ^{7}Li(p,n)^{7}Be reaction at an energy between 3 and 14 MeV must consequently be preferred. Below 11.2 MeV no nuclear interferences from beryllium, carbon and nitrogen occur. For the interference from boron a correction can be made by using the interference factor and the boron concentration determined by using the ^{10}B(d,n)^{11}C reaction (see below). Another possibility is to apply a proton and a deuteron irradiation, each followed by measurement of the induced ^{7}Be activity. Since the extent of interfer-

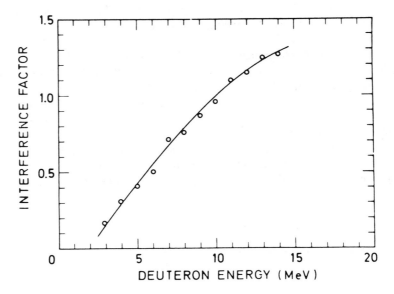

Fig. 4.10 — Interference of lithium with the determination of boron by the $^{10}B(d,\alpha n)^7Be$ reaction. (Reprinted with permission, from R. Mortier, C. Vandecasteele, K. Strijckmans and J. Hoste, *Anal. Chem.*, 1984, **56**, 2166. Copyright 1984, American Chemical Society).

ence from boron is different for the two irradiations, a simultaneous determination of boron and lithium is possible.

The nuclear reactions used to determine boron in geological materials are, of course, the same as for boron in metals and semiconductors (Table 4.1, p. 85).

As an example, the determination of lithium and boron in magnesium oxide [22] will be discussed in some detail.

4.2.1 Determination of lithium and boron in magnesium oxide [22]

Magnesium oxide containing controlled concentrations of magnesium borate is used as a coating for silicon-alloyed steel in electrical transformers. The magnesium layer protects the metal surface, is heat-resistant, and is a good electrical insulator. The determination of boron in magnesium oxide powder presents considerable difficulties. Mortier *et al.* [22] used the $^{10}B(p,\alpha)^7Be$, $^{10}B(d,\alpha n)^7Be$, and $^{10}B(d,n)^{11}C$ reactions for the determination of boron. These nuclear reactions along with the nuclear interferences have already been given in Table 4.1 (p. 85). By an appropriate choice of the incident energy (below 11.3, 14.5 and 5.8 MeV, respectively) all nuclear interferences were avoided, except the interference from lithium for the reactions yielding ^7Be. A simultaneous determination of boron and lithium was, however, possible by carrying out two irradiations, one with protons and the other with deuterons, each followed by measurement of the induced ^7Be activity.

As standards for boron, pure elemental boron and boric acid powder (H_3BO_3)

were used, for the $^{10}B(d,n)^{11}C$ reaction and for the reactions yielding ^7Be, respectively. For the irradiation of the powdered samples the sample holder shown in Fig. 3.6 (p. 59) was used. A copper or a nickel foil was placed in front of the sample or the standard to serve as a beam intensity monitor. Between the monitor foil and the sample two 5.4 mg/cm^2 aluminium foils were placed to stop recoil nuclei. Foils of the same material with the same thickness were placed before the standards. Table 4.19

Table 4.19 — Irradiation conditions for the samples.

	$^{10}B(p,\alpha)^7Be$	$^{10}B(d,\alpha n)^7Be$	$^{10}B(d,n)^{11}C$
Beam energy, MeV	6	10	7
Beam intensity, μA	2	2	2
Effective incident energy, MeV	4.5	9.0	5.8
Irradiation time, min	10–30	10–30	5–20
Monitor foil	Cu 26.8 mg/cm^2	Ni 11.1 mg/cm^2	Ni 11.1 mg/cm^2

summarizes the irradiation conditions for the samples and gives information on beam intensity monitoring. The standards were irradiated for 1 min at 50 nA intensity.

For the determination of boron by the $^{10}B(p,\alpha)^7Be$ reaction, the samples were measured, after a cooling period of 14 d, with a Ge(Li) detector. The background in the γ-ray spectrum is mainly due to the natural background and to ^{22}Na from the $^{25}Mg(p,\alpha)^{22}Na$ reaction.

Instrumental detection of ^7Be in magnesium oxide irradiated with deuterons is hindered by ^{22}Na and ^{24}Na produced by the $^{24}Mg(d,\alpha)^{22}Na$, $^{25}Mg(d,\alpha n)^{22}Na$, and $^{26}Mg(d,\alpha)^{24}Na$ reactions. Sodium was therefore separated by fixation on a column filled with hydrated antimony pentoxide. In order to measure the standard and the sample under the same geometrical conditions, beryllium was precipitated from the eluate as the hydroxide, which was ignited to beryllium oxide and measured with a Ge(Li) detector.

In the determination of boron by using the $^{10}B(d,n)^{11}C$ reaction, ^{11}C cannot be detected instrumentally because of the activity formed from the matrix. Therefore, ^{11}C is isolated as carbon dioxide by dissolving the sample in an oxidizing alkaline melt. The cooled melt is dissolved in acid and the carbon dioxide liberated is absorbed in a sodium hydroxide solution, and the annihilation radiation is measured with a γ-γ coincidence set-up.

From the apparent boron concentrations (i.e. the concentrations obtained when all the ^7Be activity is attributed to boron) $c_{B,p}$ obtained by proton activation and $c_{B,d}$ by deuteron activation, the boron c_B, and lithium, c_{Li}, concentrations can be calculated by solving the equations

$$c_{B,p} = c_B + IF_p \, c_{Li} \qquad (4.1)$$

$$c_{B,d} = c_B + IF_d \, c_{Li} \qquad (4.2)$$

Table 4.20 gives the boron and lithium concentrations calculated by means of Eqs. (4.1) and (4.2), and the boron concentration obtained by the $^{10}B(d,n)^{11}C$ reaction. For the results obtained from the ^{7}Be activity the standard deviation for the lithium and boron concentrations was calculated from the set of equations obtained by application of the formulae for the propagation of random error to Eqs. (4.1) and (4.2). The lithium concentrations for all samples have large standard deviations, indicating that the lithium is not homogeneously distributed. The results for boron by using ^{7}Be and ^{11}C have an acceptable precision (4–10%), except for material 4, which was a blank. For none of the materials analysed is there a significant difference between the results for boron obtained by the two methods.

Table 4.20 — Results ($\mu g/g$) for lithium and boron in magnesium oxide.

Material		^{7}Be		^{11}C
		Boron	Lithium	Boron
1	$\bar{x}$	854	26	838
	s	47	16	58
	n	3	3	4
2	$\bar{x}$	80.2	1.8	87.0
	s	8.3	2.0	8.9
	n	3	3	4
3	$\bar{x}$	14.79	0.84	14.1
	s	0.67	0.21	1.1
	n	3	3	2
4	$\bar{x}$	1.51	0.98	1.88
	s	0.37	0.17	0.27
	n	3	3	4

4.3   CONCLUSION

CPAA allows the determination of lithium, boron, carbon, nitrogen and oxygen in almost every type of solid material. In many cases, a chemical separation is required to separate the radionuclide of interest from the matrix. Intercomparisons have shown that CPAA is, in general, one of the most accurate methods available for these elements. The detection limits are usually in the 1–100 ng/g range, depending on the nuclear reaction, the irradiation conditions and the matrix.

REFERENCES

[1] C. Engelmann, G. Kraft, J. Pauwels and C. Vandecasteele, *Modern Methods for the Determination of Non-Metals in Non-Ferrous Metals*, de Gruyter, Berlin, 1985.
[2] A. Giovagnoli, M. Valladon, C. Koemmerer, G. Blondiaux and J. L. Debrun, *Anal. Chim. Acta*, 1979, **109**, 411.
[3] K. Strijckmans, C. Vandecasteele and M. Esprit, *Z. Anal. Chim.*, 1980, **303**, 106.
[4] P. Goethals, C. Vandecasteele and J. Hoste, *Anal. Chim. Acta*, 1979, **108**, 367.
[5] J. Petit, J. Gosset and Ch. Engelmann, *J. Radioanal. Chem.*, 1980, **55**, 69.

[6] C. Vandecasteele, K. Strijckmans, Ch. Engelmann and H. M. Ortner, *Talanta*, 1981, **28**, 19.
[7] R. Mortier, C. Vandecasteele, K. Strijckmans and J. Hoste, *Anal. Chem.*, 1984, **56**, 2166.
[8] C. Vandecasteele and J. Hoste, *Anal. Chim. Acta*, 1975, **79**, 302.
[9] B. Vialatte, J. N. Barrandon, S. Alexandrov, I. N. Bourelly, C. Cleyrergue, N. Deschamps and H.Jaffrezic, *Radiochem. Radioanal. Lett.*, 1970, **5**, 59.
[10] G. Blondiaux and B. Vialatte, *Radiochem. Radioanal. Lett.*, 1971, **8**, 251.
[11] A. Kohn, J. N. Barrandon, J. L. Debrun, M. Valladon and B. Vialette, *Anal. Chem.*, 1974, **46**, 1737.
[12] N. Jaffrezic-Renault, *Radiochem. Radioanal. Lett.*, 1977, **29**, 47.
[13] M. Fedoroff, L. Debove and C. Loos-Neskovic, *J. Radional. Chem.*, 1974, **21**, 331.
[14] L. Faure, M. Boissier and J. Tousset, *J. Radioanal. Chem.*, 1972, **10**, 213.
[15] L. Faure, J. Giroux and J. Tousset, *J. Radioanal. Chem.*, 1972, **10**, 223.
[16] H. Chermette, C. Martelet, D. Sandino, M. Benmalek and J. Tousset, *Anal. Chim. Acta*, 1972, **59**, 373.
[17] T. Nozaki, *J. Radioanal. Chem.*, 1982, **72**, 526.
[18] Ch. Engelmann, *J. Radioanal. Chem.*, 1971, **7**, 89.
[19] Ch. Engelmann, *J. Radioanal. Chem.*, 1971, **7**, 281.
[20] P. N. Kuin, *Report* EUR 3896 d-f-e, 1968.
[21] H. Rommel, *Anal. Chim. Acta*, 1966, **34**, 427.
[22] R. Mortier, C. Vandecasteele, J. Hoste and F. den Hartog, *J. Radional. Nucl. Chem. Lett.*, 1986, **1**, 47.
[23] R. Mortier, *Ph.D. Thesis*, Ghent University, Belgium, 1983.
[24] Ch. Engelmann, J. Gosset and J. M. Rigaud, *Radiochem. Radioanal. Lett.*, 1970, **5**, 319.
[25] S. Shibata, S. Tanaka, T. Suzuki, H. Umezawa, J. G. Lo and S. J. Yeh, *Int. J. Appl. Radiat. Isot.*, 1979, **30**, 563.
[26] P. Goethals, C. Vandecasteele and J. Hoste, in *Analysis of Non-Metals in Metals*, G. Kraft (ed.), p. 259. De Gruyter, Berlin, 1980.
[27] R. Mortier, C. Vandecasteele and J. Hoste, *Anal. Chim. Acta*, 1980, **121**, 147.
[28] R. Mortier, K. Strijckmans and C. Vandecasteele, *Bull. Soc. Chim. Belg.*, 1981, **90**, 297.
[29] R. Mortier, C. Vandecasteele and J. Hoste, in *Analysis of Non-Metals in Metals*, G. Kraft (ed.), p. 267. De Gruyter, Berlin, 1980.
[30] C. S. Sastri, R. Caletka and V. Krivan, *Anal. Chem.*, 1981, **53**, 765.
[31] Ph. Albert, G. Chaudron and P. Süe, *Bull. Soc. Chim. France*, 1953, **20**, C97.
[32] Ph. Albert, A. Nouaille, G. Chaudron and P. Süe, *Congr. Int. de l'Aluminium*, 14–19 June 1954, p. 191. Documentation des Alliages légers, Paris.
[33] T. Nozaki, Y. Yatsurugi and N. Akiyama, *J. Radioanal. Chem.*, 1970, **4**, 87.
[34] Ch. Engelmann and A. Marschal, *Radiochem. Radioanal. Lett.*, 1971, **6**, 189.
[35] Y. Endo, Y. Yatsurugi, N. Akiyama and T. Nozaki, *Anal. Chem.*, 1972, **44**, 2258.
[36] F. Mayolet, P. Reimers and Ch. Engelmann, *J. Radioanal. Chem.*, 1972, **12**, 115.
[37] J. Martin and E. Haas, *Z. Anal. Chim.*, 1972, **259**, 97.
[38] T. Nozaki, Y. Yatsurugi, N. Akiyama, Y. Endo and Y. Makide, *J. Radioanal. Chem.*, 1974, **19**, 109.
[39] C. Vandecasteele, F. Adams and J. Hoste, *Anal. Chim. Acta*, 1974, **72**, 269.
[40] T. Nozaki, Y. Yatsurugi and Y. Endo, *J. Radioanal. Chem.*, 1976, **32**, 43.
[41] C. Vandecasteele, K. Strijckmans and J. Hoste, *Anal. Chim. Acta*, 1979, **108**, 127.
[42] M. Valladon, G. Blondiaux, C. Koemmerer, J. Hallais, G. Poiblaud, A. Huber and J. L. Debrun, *J. Radioanal. Chem.*, 1980, **58**, 169.
[43] A. O. Sanni, N. G. Roche, H. J. Dowell, E. A. Schweikert and T. H. Ramsey, *J. Radioanal. Nucl. Chem.*, 1984, **81**, 125
[44] C. Vandecasteele, J. Dewaele, J. Hoste, R. De Doncker, F. Vangaever and J. Vanhumbeek, *J. Radioanal. Nucl. Chem. Lett.*, 1985, **95**, 167.
[45] M. L. Böttger, D. Birnstein, W. Helbig and S. Niese, *J. Radioanal. Chem.*, 1980, **58**, 173.
[46] H. M. Ortner, *Talanta*, 1979, **26**, 629.
[47] C. Vandecasteele, *The Certification of Carbon and Nitrogen in Molybdenum*, *Report* EUR 10213 EN, 1985.
[48] T. Nozaki, T. Okuo, H. Akutzu and M. Furukawa, *Bull. Chem. Soc. Japan*, 1966, **39**, 2685.
[49] K. Strijckmans, C. Vandecasteele and J. Hoste, *Anal. Chim. Acta*, 1977, **89**, 255.
[50] K. Strijckmans, C. Vandecasteele and J. Hoste, *Anal. Chim. Acta*, 1978, **96**, 195.
[51] J. Petit, J. Gosset and Ch. Engelmann, *Mém. Sci. Rev. Met.*, 1978, **75**, 395.
[52] A. Giovagnoli, M. Valladon, C. Koemmerer, G. Blondiaux and J. L. Debrun, *Anal. Chim. Acta*, 1979, **109**, 411.
[53] C. Vandecasteele, K. Strijckmans, R. Kieffer and J. Hoste, *Radiochem. Radioanal. Lett.*, 1979, **38**, 261.
[54] C. S. Sastri and V. Krivan, *Anal. Chem.*, 1981, **53**, 2242.
[55] K. Strijckmans, R. Mortier, C. Vandecasteele and J. Hoste, *Mikrochim. Acta*, 1982 **II**, 321.

[56] J. Pauwels and L. Haemers, *The Certification of Nitrogen in Non-ferrous Metals. Nitrogen in Zirconium* (BCR NO 21-56-57), *Report* EUR 6939EN, 1980.
[57] K. Saito, T. Nozaki, S. Tanaka, M. Furukawa and H. Cheng, *Int. J. Appl. Radiat. Isot.*, 1963, **14**, 357.
[58] G. Revel and Ph. Albert, *J. Nucl. Mater.*, 1968, **25**, 87.
[59] J. L. Debrun, J. N. Barrandon and Ph. Albert, *Bull. Soc. Chim. France*, 1969, 1011.
[60] E. A. Schweikert and H. L. Rook, *Anal. Chem.*, 1970, **42**, 1525.
[61] D. M. Lee, C. V. Stauffacher and S. S. Markowitz, *Anal. Chem.*, 1970, **42**, 994.
[62] Ch. Engelmann, *J. Radional. Chem.*, 1970, **6**, 227.
[63] C. Kim, *Anal. Chim. Acta*, 1971, **54**, 407.
[64] B. Vialatte, *J. Radioanal. Chem.*, 1973, **17**, 301.
[65] C. Vandecasteele, F. Adams and J. Hoste, *Anal. Chim. Acta*, 1974, **71**, 67.
[66] M. Fedoroff, L. Debove and C. Loos-Neskovic, *J. Radioanal. Chem.*, 1974, **21**, 331.
[67] E. A. Schweikert, J. R. McGinley, G. Francis and D. L. Swindle, *J. Radioanal. Chem.*, 1974, **19**, 89.
[68] C. Vandecasteele and J. Hoste, *Anal. Chim. Acta*, 1975, **78**, 121.
[69] C. Vandecasteele, F. Adams and J. Hoste, *Anal. Chim. Acta*, 1975, **76**, 27.
[70] M. Valladon and J. L. Debrun, *J. Radioanal. Chem.*, 1977, **39**, 385.
[71] C. Vandecasteele and J. Hoste, *J. Radioanal. Chem.*, 1975, **27**, 465.
[72] C. Vandecasteele, P. Goethals, R. Kieffer and J. Hoste, *Bull. Soc. Chim. Belg.*, 1975, **84**, 673.
[73] H. Petri and C. S. Sastri, *Z. Anal. Chim.*, 1975, **277**, 25.
[74] M. Valladon, A. Giovagnoli and J. L. Debrun, *Analusis*, 1978, **6**, 452.
[75] P. Debefve, P. Lerch and C. Vandecasteele, *Radiochem. Radioanal. Lett.*, 1976, **24**, 51.
[76] T. Nozaki, M. Iwamoto, K. Usami, K. Mukai and A. Hiraiwa, *J. Radioanal. Chem.*, 1979, **52**, 449.
[77] H. Petri and G. Erdtmann, in *Analysis of Non-Metals in Metals*, G. Kraft (ed.), p. 253. De Gruyter, Berlin, 1981,
[78] M. Fedoroff, C. Loos-Neskovic, J. C. Rouchaud and G. Revel, in *Analysis of Non-Metals in Metals*, G. Kraft (ed.), p. 243. De Gruyter, Berlin, 1981.
[79] J. L. Debrun and J. N. Barrandon, *Proc. 7th Int. Conf. on Cyclotrons and their Applications*, p. 507. Birkhäuser, 1975.
[80] F. Dugain, M. André and A. Speecke, *Radiochem. Radioanal. Lett.*, 1970, **4**, 35.
[81] F. Dugain and C. Michaut, *Radiochem. Radioanal. Lett.*, 1972, **9**, 119
[82] J. Pauwels, *The Certification of Oxygen in Nonferrous Metals. Oxygen in Primary Ingot Aluminium* (BCR No 25), *Report* EUR 6240 EN, 1979.
[83] R. Mortier, C. Vandecasteele, J. Hertogen and J. Hoste, *J. Radioanal. Chem.*, 1982, **71**, 189.
[84] B. Vialatte, *J. Radioanal. Chem.*, 1971. **8**, 269.
[85] C. S. Sastri, R. Caletka and V. Krivan, *Anal. Chem.*, 1981, **53**, 765.

5

Determination of medium and heavy elements in metals and semiconductors

The analysis of metals and semiconductors for traces of medium and heavy elements is another main application of CPAA. For this application, important advantages of CPAA relative to other techniques are:

— high sensitivity;
— high accuracy, with no need for standards with the same bulk composition as the samples;
— instrumental determination of several elements in a number of matrices;
— no risk of contamination with inactive material, if a surface layer is removed after the irradiation.

In recent years, several analytical techniques with very low detection limits have been developed. Techniques such as electrothermal atomization–atomic-absorption spectrometry (ETA–AAS), inductively-coupled plasma atomic-emission spectrometry (ICP–AES), and inductively-coupled plasma mass-spectrometry (ICP–MS) require, for the accurate analysis of solid materials, previous dissolution of the sample, with the inherent risk of contamination and introduction of reagent blanks, often limiting the detection limits achieved in practice. Techniques such as spark-source mass-spectrometry or glow-discharge mass-spectrometry, that do not require previous dissolution of solid samples and are ideally suited for the analysis of metals and semiconductors, have most of the advantages mentioned for CPAA. For accurate calibration, however, these techniques usually rely on standards consisting of the matrix of interest, analysed by other analytical techniques.

Although all light projectiles (p, d, t, ^{3}He, ^{4}He) have been applied for the determination of medium and heavy elements in metals and semiconductors, protons are most often used. Proton activation analysis will therefore be treated separately and in more detail.

5.1 PROTON ACTIVATION ANALYSIS

5.1.1 Nuclear reactions, detection limits

For the determination of medium and heavy elements by CPAA, protons have most often been applied as projectiles, since they allow the sensitive interference-free and purely instrumental determination of many elements in a number of matrices. The incident proton energy chosen depends on several factors. When a purely instrumental analysis is to be made, both the activity induced in the matrix and the occurrence of nuclear interferences must be considered. For many matrices an incident energy of around 10–12 MeV is optimal. When chemical separations are used to separate the indicator radionuclide(s) of interest from other radionuclides produced from the matrix, the incident energy may often be higher, since it is not necessary to avoid activation of the matrix.

Extensive systematic studies of activation analysis with 10 MeV protons combined with γ-ray spectrometry with a germanium semiconductor detector have been made by Debrun et al. [1] and by Barrandon et al. [2]. The studies were limited to radionuclides with a half-life longer than 15 min. Thirty-nine elements with atomic number between 16 and 82, as the pure element whenever possible, or as a compound with accurately known composition, were irradiated with 10 MeV protons. Up to 12 different targets were mounted on a rotary irradiation set-up and irradiated for 5–20 min at a total extracted current of ca. 1 μA, so the actual average beam intensity on each target was ca. 0.05 μA. The beam intensity was deduced from the ^{55}Co, ^{52m}Mn and ^{52}Mn activities induced in a 10-μm thick Havar foil placed before the irradiated samples. The relation between the induced activity and the beam intensity was determined experimentally by irradiation of a Havar foil with measurement of the incident current with a Faraday cup.

One or several radionuclides are produced by (p,n) reactions of the element considered and each radionuclide emits one or several γ-rays. For each element studied, the induced activity of all radionuclides of interest was measured with a germanium semiconductor detector.

The results are summarized in Table 5.1. From the experimental number of counts for a given γ-ray peak, the count-rate at the end of the irradiation was calculated by using Eq. (2.50). Correction for the detector efficiency yielded the number of γ-rays emitted per min, which was normalized to correspond to 1 μg/g of the element, an irradiation time of 1 hr, and a beam intensity of 1 μA, by using Eqs. (2.46) and (2.47). The values, N_γ, were normalized to the range of 10 MeV protons in aluminium by using Eq. (5.1):

$$(N_\gamma)_{Al} = N_\gamma \frac{R_{Al}}{R} \tag{5.1}$$

where R_{Al} and R are the ranges in aluminium and in the irradiated material. $(N_\gamma)_{Al}$ is given in column 6 of Table 5.1.

From these data detection limits were calculated on the assumptions that the minimum detectable activity for a given peak is 3 times the square root of the background activity, the samples are measured (directly after irradiation) for 1.8 $t_{1/2}$ or 60 hr, whichever is the shorter, with a germanium semiconductor detector with a relative detection efficiency of 22%, and a resolution of 2.7 keV, and that the samples are irradiated for 1 hr at 1 μA intensity.

The detection limits are given in column 7 of Table 5.1. It appears that for the experimental conditions described, 27 elements have a detection limit below 50 ng/g. Of course, the detection limits given in Table 5.1 are lower than the actual detection limits obtained when a real sample is analysed, because of the activity induced in the matrix and in the other impurities. For the instrumental analysis of silicon, for example, the main matrix activity is due to ^{30}P from the $^{30}Si(p,n)^{30}P$ reaction. Figure 5.1 compares the calculated (Table 5.1) and experimental detection limits for a silicon matrix, using for all the elements the most sensitive radionuclide and γ-ray energy. It appears that in general, the detection limits for the heavy elements ($Z>60$) are higher than those for the other elements. This is due to the influence of the Coulomb barrier, which increases with atomic number, as shown in Fig. 2.2 (p. 16), and is about 10 MeV for $Z=60$. In order to lower considerably the detection limits for the heavy elements, it may therefore be useful to choose a higher energy.

Table 5.1 — Induced activity and detection limits for irradiation with 10 MeV protons [1, 2]. (Reprinted with permission, from J. L. Debrun, J. N. Barrandon and P. Benaben, *Anal. Chem.*, 1976, **48**, 167, copyright 1976, American Chemical Society, and from *Anal. Chim. Acta*, 1976, **83**, 157, copyright 1976, Elsevier Science Publishers B.V.)

Element	Target	Radionuclide	Half-life, hr	γ-Energy, keV	Number of γ/min normalized to Al,* $(N_\gamma)_{Al}$	Calculated detection limit, ng/g†
S	S	^{34m}Cl	0.53	146	257.5	21.5
				1106.5	57	
				1177.5	8.4	
Ca	Glass	^{44}Sc	3.95	1157	688	8.8
		^{44m}Sc	58.6	271.2	1.8	3190
		^{44}Sc	43.7	983.4	11.5	165
				1037.4	11.5	175
				1311.6	11.2	200
Ti	Ti	^{48}V	386.4	944.2	15.5	—
				983.5	196	9
				1311.8	196	9
V	V	^{51}Cr	667.7	320	23.5	43
Cr	Cr	^{52m}Mn	0.35	1434.4	40650	0.4
		^{52}Mn	134.4	744.1	50.5	28
				935.5	54.5	31
				1434.2	61	22
		^{54}Mn	7500	834.7	1.2	1225
Fe	Fe	^{56}Co	1855.2	846.7	25.5	61
				1037.6		
				1238	3.3	
				1771.4	4.5	
		^{57}Co	6480	122.1	1.2	
Ni	Ni	^{60}Cu	0.39	826.2	1120	
				1332.5	5710	5.5
				1792	3440	4.5
		^{61}Cu	3.41	283	36	86
				656.3	35	130
				1185.7	13.5	490
		^{55}Co	17.9	931.5	16	150

Continued next page

Table 5.1 (contd.)

Element	Target	Radionuclide	Half-life, hr	γ-Energy, keV	Number of γ/min normalized to Al,* $(N_\gamma)_{Al}$	Calculated detection limit, ng/g†
Cu	Cu	^{63}Zn	0.64	669.8	3600	3
				962.2	3110	4
		^{65}Zn	5851.2	1115.5	5.3	390
Zn	Zn	^{66}Ga	9.4	834	83	
				1039.3	650	6
				1333.8	15.5	
				1730	36	
				1918	30	
		^{67}Ga	78.1	93.1	26.5	40
				184.2	13.8	60
				300	10.8	90
		^{68}Ga	1.14	1076.8	434.8	35
Ga	GaAs	^{69}Ge	39	553.4	9	
				574.1	166	
				872.4	180	
				1106.6	500	4
				1336.6	58	
Ge	Ge	^{70}As	0.87	744.9	620	
				1040.3	2720	5
				1114.6	570	
		^{72}As	26	630	51.5	
				834	620	3
		^{74}As	424.8	595.9	62	22
				634.9	17	
		^{76}As	26.3	559.2	146	10
				657.2	17.5	
As	Ga As	^{75}Se	2880	121.2	6.5	
				136.1	21	40
				264.4	19.5	40
				279.4	9.5	
				400.5	9.5	
Se	Se	^{76}Br	15.9	559.3	180	10
		^{77}Br	56	520.9	32.8	42
		^{80}Br	0.3	665.7	10600	1.5
		^{82}Br	35.4	554.3	124	
				619.1	100	
				698.4	51	10
				776.6	77	
				827.8	58	
Br	KBr	^{79}Kr	34.9	217.5	25.5	
				261.4	160	6
				306.8	29	
				397.6	120	9
				606	115	12
Rb	RbCl	^{85m}Sr	1.17	151.3	1260	0.7
				231.7	8450	
		^{87m}Sr	2.83	388.5	2120	2
Sr	SrHPO$_4$	^{84}Y	0.64	793	28	410
				974.3	21.5	620
				1039.7	14.5	1000
		^{86}Y	14.6	627.8	73	
				777.5	50	
				1077.4	210	13
				1153.4	70	
				1920	72.5	

Continued next page

Table 5.1 (*contd.*)

Element	Target	Radionuclide	Half-life, hr	γ-Energy, keV	Number of γ/min normalized to Al,* $(N_\gamma)_{Al}$	Calculated detection limit, ng/g†
Sr	SrHPO$_4$	^{87}Y	80	484.9	27.5	44
		^{88}Y	2568	898.2	26	66
				1835.8	26	
Y	Y	^{89}Zr	78.4	909.1	1100	2
Zr	Zr	^{90}Nb	14.6	141.2	640	2
				1129.1	870	3.5
		^{92m}Nb	243.8	934.5	51.5	34
		^{95}Nb	842.4	765.6	0.7	2350
		^{96}Nb	23.35	460.2	18	
				569	35	
				778.7	63	30
Nb	Nb	^{93m}Mo	6.9	263.1	46	44
				684.8	95	31
				1477.1	110	30
Mo	Mo	^{94}Tc	4.9	702.9	83	45
				871.3	95	45
		^{95}Tc	20	765.8	380	5
				1074	15.5	
		^{96}Tc	104.4	314	1.7	
				778.2	88	18
				812.6	68	22
				1127	13	
		^{99m}Tc	6.05	140.5	65	30
Ru	Ru	^{96}Rh	0.155	630.7	290	
				684.6	440	49
		^{98}Rh	0.150	651.8	645	32
		^{99m}Rh	4.7	340.5	510	5.5
				618.2	106	
				1260.7	109	
		^{99}Rh	386.4	352.4	1.5	720
				528.2	2	628
		^{100}Rh	20	446.0	28	
				539.7	206	8
				822.5	50	
				1361.8	51	
				1553.4	52	
		^{101}Rh	28908	197.8	0.06	15400
		^{101m}Rh	103.2	306.8	40	20
Rh	Rh	^{103}Pd	420	294.4	0.004	
				357.0	0.033	31500
				497.0	0.006	
Pd	Pd	^{104}Ag	1.10	555.8	1270	5.5
				767.4	775	11
				785.5	116	
				923.2	210	
				941.4	280	34
		^{105}Ag	960	280.3	2.7	340
				344.2	4	265
				443.4	0.9	
				1088.7	0.56	
		^{106m}Ag	201.6	220.9	0.09	
				450.6	1	1150
				717.3	1.3	1200
				748.1	0.7	
				1045.0	1.40	1480

Continued next page

Table 5.1 (*contd.*)

Element	Target	Radionuclide	Half-life, hr	γ-Energy, keV	Number of γ/min normalized to Al,* $(N_\gamma)_{Al}$	Calculated detection limit, ng/g†
Pd	Pd	^{110m}Ag	6240	657.7	0.30	4850
				763.9	0.08	
				884.7	0.29	6100
				937.5	0.12	
				1384.2	0.1	
Ag	Ag	^{107}Cd	6.5	93	119	17.2
				796.2	2	
				829	5	750
Cd	Cd	^{110m}In	4.9	641.7	8.3	
				657.7	27.3	135
				707.4	9.8	
				884.7	28	
				937.5	21.5	
		^{111}In	67.44	171.4	66	12.5
				245.4	65	12.5
		^{113m}In	1.66	391.7	608	8
		^{114m}In	1200	189.8	0.76	1150
				558.4	0.15	8500
				725.4	0.13	1100
		^{115m}In	4.4	336.2	20	125
In	InP	^{113}Sn	2760	255	0.011	
				391.7	0.35	3000
Sn	Sn	^{116}Sb	0.267	933	975	20
		^{117}Sb	2.8	158.7	380	7.5
		^{118m}Sb	5.1	253.7	23	100
				1051	29	180
				1229.6	27.55	200
		^{120m}Sb	139.2	197.1	1.4	660
				1022.8	1.35	1450
		^{122}Sb	67.2	564	18.5	70
Sb	Sb	^{121}Te	408	507.6	5.9	
				573.2	27	40
		^{121m}Te	3600	212.1	1.4	580
		^{123m}Te	2808	159.1	1.9	650
Te	Te	^{123}I	13.3	159	18.6	70
		^{124}I	100	602.9	11	125
		^{126}I	312	388.5	6	180
				666.3	6.4	225
		^{128}I	0.417	442.9	1470	7
				526.7	104	
		^{130}I	12.3	418	113	
				536	390	5.4
				668.5	393	6
				739.5	325	
				1157.2	62	
I	KI	^{127}Xe	873.6	145.2	1.1	
				172.1	6.4	
				202.8	20	42
				375	6.3	
Ba	BaCl$_2$	^{135}La	19.5	480.5	0.7	2300
W	W	^{182}Re	64	169.3	0.07	
				191.6	0.6	

Continued next page

Table 5.1 (*contd.*)

Element	Target	Radionuclide	Half-life, hr	γ-Energy, keV	Number of γ/min normalized to Al,* $(N_\gamma)_{Al}$	Calculated detection limit, ng/g†
W	W	^{182}Re	64	276.4	0.096	10800
				286.6	0.06	
				351.2	0.09	12200
		^{183}Re	1680	162.4	0.14	6000
Re	Re	^{185}Os	2246.4	645.9	1.6	840
				717.1	0.09	
				874.6	0.12	
				880	0.1	
Ir	Ir	^{191}Pt	68.4	172.2	0.33	
				351.2	0.44	
				359.8	0.6	
				409.4	0.84	
				538.8	1.9	600
Pt	Pt	^{192}Au	4.1	316.3	1.6	1870
		^{194}Au	39.5	293.5	1.9	
				328.3	1.5	70
		^{196}Au	148.8	333.0	0.80	
				355.7	4.3	250
				426.0	0.34	
		^{198}Au	64.6	411.8	2.8	375
Au	Au	^{197m}Hg	24	134.0	6.7	140
				278.9	0.8	
		^{197}Hg	64.1	191.4	0.2	4900
Hg	HgCl$_2$	^{198}Tl	5.3	411.8	20	130
		^{199}Tl	7.4	158.2	1.7	
				208.1	4.4	430
				247.0	3	630
				455.1	5.5	440
		^{200}Tl	26.1	368.0	13.7	95
				579.6	1.7	
				828.3	1.2	
		^{202}Tl	288	439.6	2.0	
				520.2	0.03	
Tl	Tl	^{203}Pb	52.1	279.2	8.80	103
Pb	Pb	^{206}Bi	149.8	343.5	0.52	
				515.8	1.1	
				802.8	2.9	520
				880.7	2.2	
				1718.7		

*Number of γ-rays emitted per min for a thick aluminium target containing 1 µg/g of the element of interest, at the end of a 1 hr irradiation with 10 MeV protons at 1 µA beam intensity.
†The detection limit is calculated for a minimum detectable activity of 3 times the square root of the background, and for an aluminium sample.

Similar information concerning nuclear reactions yielding radionuclides with a half-life longer than 15 min was compiled by Barrandon *et al.* [3], who studied nuclear reactions induced by 11 MeV protons for 30 elements, by Kormali and Schweikert [4], who surveyed the application of proton activation analysis for elements with $40 < Z < 74$, and by Krivan and Krivan [5], who measured thick target

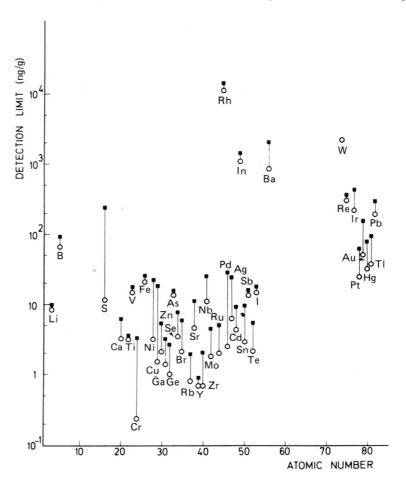

Fig. 5.1 — Calculated (○) and experimental (□) detection limits for silicon.

yields and analytical sensitivities for nuclear reactions induced with 12 MeV protons for titanium, iron, zinc, germanium, zirconium, niobium, palladium, tantalum, and tungsten. Barrandon *et al.* [6] give the excitation functions of the most important proton-induced nuclear reactions of titanium, vanadium, chromium, iron, nickel, copper and zinc, whereas Kormali *et al.* [7] give the excitation functions for proton energies up to 20 MeV for some important nuclear reactions of copper, tin, molybdenum, tellurium, palladium, cadmium and caesium. These excitation functions are, of course, very useful for the selection of the optimum incident energy for a given analysis. They can also be used in calculations of the *F*-factor (Section 3.4.3).

Debrun *et al.* [8] also studied purely instrumental proton activation analysis based on short-lived radionuclides, i.e. with a half-life ranging from 1 sec to 1 min. Only nuclear reactions yielding a radionuclide that emits one or more γ-rays were considered. Proton activation producing short-lived radionuclides was found suitable for the determination of traces of selenium, bromine, yttrium, zirconium, lanthanum, praseodymium, neodymium and dysprosium. Table 5.2 gives the nuclear reactions and the thick target yields for protons of 16.8 MeV. These yields can in practice be increased by a factor of 20–150, depending on the half-lives of the nuclides considered, when the intensity is increased to 10 μA and the irradiation time to 15 sec. The elements mentioned can be determined in 13 matrix elements that yield little or no activity under the experimental conditions described, viz. lithium, beryllium, vanadium, manganese, cobalt, gallium, arsenic, antimony, caesium, europium, terbium, holmium, and bismuth. Of these, only beryllium, manganese and bismuth may be irradiated at 16.8 MeV. For the other elements a lower proton energy must be chosen. It is thus clear that proton activation analysis based on short-lived radionuclides is limited in its scope to a rather small set of trace elements and matrix combinations.

Table 5.2 — Thick target yields for some nuclear reactions yielding short-lived radionuclides*. (Reprinted with permission, from J. L. Debrun, D. C. Riddle and E. A. Schweikert, *Anal. Chem.*, 1972, **44**, 1386. Copyright 1972 American Chemical Society)

Element	Target	Reaction	Yield*, counts/μA
Se	Se	^{80}Se(p, 2n)^{79m}Br	6×10^6
Br	KBr	^{79}Br(p, n)^{79m}Kr	33×10^6
		^{79}Br(p, p′)^{79m}Br	2.2×10^6
		^{81}Br(p, n)^{81m}Kr	16×10^6
Y	Y$_2$O$_3$	^{89}Y(p, p′)^{89m}Y	0.8×10^6
		^{89}Y(p, n)^{89m}Zr	10×10^6
Zr	Zr	^{90}Zr(p, n)^{90m}Nb	23×10^6
La	La$_2$O$_3$	^{139}La(p, n)^{139m}Ce	7.5×10^6
Pr	Pr$_2$O$_3$	^{141}Pr(p, n)^{141m}Nd	7×10^6
Nd	Nd$_2$O$_3$	^{142}Nd(p, n)^{142}Pm	20×10^6
Dy	Dy$_2$O$_3$	^{161}Dy(p, n)^{161m}Ho	2.8×10^6
		^{163}Dy(p, n)^{163m}Ho	0.8×10^6

*Experimental conditions: the samples are irradiated with 16.8 MeV protons for 1 sec at 1 μA intensity and measured at a distance of 5 cm from a 7.5 × 7.5 cm NaI(Tl) scintillation detector; the yield corresponds to the end of the irradiation for a counting time long enough to ensure complete disintegration of the radionuclide of interest; the count rates were corrected to correspond to 100% of the element of interest.

McGinley and Schweikert [9] studied energy-dispersive X-ray counting of radio-nuclides produced by proton or deuteron activation for the instrumental determination of trace elements with $Z \geqslant 26$. Experimentally determined thick target yields and detection limits for 37 elements were presented [9]. The detection limits ranged from 10^{-4} μg for molybdenum to 3.4 μg for rhodium, for a 3 μA irradiation with 20 MeV protons or deuterons for 3 hr or 1 half-life of the product radionuclide, whichever is the shorter, and a counting time of 1 half-life. This method may suffer from systematic errors due to self-absorption of the X-rays. The degree of self-absorption is a function of the X-ray energy, of the matrix material, and of the depth in the sample from which the X-ray is emitted. The most suitable application is the determination of medium and high Z elements in samples composed of low Z elements ($Z < 14$).

5.1.2.   Interferences

Several sorts of nuclear interferences occur in proton activation analysis.

Primary nuclear interference occurs when a (p, n) reaction of the analyte element yields the same radionuclide as a (p, α), $(p, \alpha n)$, etc. reaction of another (interfering) element. Some examples are given in Table 5.10 (p. 129). Such nuclear interferences can often be avoided, or at least reduced, by an appropriate choice of the incident energy. When the interfering reactions cannot be avoided, corrections may sometimes be applied by using an experimentally determined interference factor (Section 2.4).

Secondary interferences are more difficult to correct for. The most common example of secondary interference is that due to neutrons produced when the target is irradiated with charged particles. The neutrons thus produced can induce nuclear reactions of the (n, p), (n, α) and $(n, 2n)$ type, and, after thermalization, also of the (n, γ) type. The neutron flux thus obtained is from one to several orders of magnitude lower than the charged particle flux. Therefore, secondary nuclear interferences are usually only important when the analyte element is present at trace concentrations and the interfering element is the matrix or a major element. Some examples of secondary interferences are:

		Example
$^{25}Mg(\alpha, p)^{28}Al$	$^{28}Si(n, p)^{28}Al$	Determination of magnesium in cast iron [10]
$^{65}Cu(p, n)^{65}Zn$	$^{64}Zn(n, \gamma)^{65}Zn$	Determination of copper in zinc [11]

Estimation of the extent of nuclear interference due to secondary neutrons is very difficult. However, to minimize the occurrence of these interferences, the use of unnecessarily thick samples should be avoided. Reactions with secondary neutrons may sometimes also result in higher detection limits than expected. Irradiation of aluminium leads, for instance, through (p, n) reaction only to ^{27}Si ($t_{1/2} = 4.16$ sec). The analysis of aluminium is, however, also disturbed by ^{28}Al from $^{27}Al(n, \gamma)^{28}Al$, by ^{27}Mg from $^{27}Al(n, p)^{27}Mg$ and by ^{24}Na from $^{27}Al(n, \alpha)^{24}Na$.

Of course, spectrometric interferences may also occur.

5.1.3   Applications

Table 5.3 reviews the literature on proton activation analysis for medium and heavy elements in metals and semiconductors. It appears that up to 1979 applications were almost exclusively limited to matrices that are activated to only a limited extent on irradiation with protons, or that yield only short-lived radionuclides: aluminium, cobalt, silver, tantalum, niobium, rhodium, gold, dysprosium, holmium, iridium, praseodymium, silicon and terbium. More recently, materials such as copper, tungsten, iron and zinc have also been analysed. These materials become highly radioactive on irradiation with protons, so radiochemical separations are required to separate the radionuclide of interest from the sample matrix.

Table 5.3 — Applications of proton activation analysis to the determination of medium and heavy elements in metals and semiconductors

Reference	Matrix analysed	Elements detected	Concentration range, $\mu g/g$	Chemical separation
[12] Debrun and Albert	Al	S		
[13] Dabney et al.	Fe, Al	S	0.3–30	Yes
[14] Riddle and Schweikert	Glass	Tl, Pb, Bi	0.1–60	Yes
	Au			
[15] Debrun and Barrandon	Al, Co, Ag	S, Ti, V, Cr, Fe, Ni, Cu, Zn, Ga, Ge, As, Se, Rb, Sr, Zr, Nb, Mo, Ru, Pd, Cd, Sn, Sb, Te, I, Ir, Pt, Au, Tl, Pb	0.002–8000	No
[16] Krivan et al.	Ta	Ti, Fe, Zr, Nb, Mo, W	0.09–33	No
[17] Barrandon et al.	Nb	Mo, W	70	No
	Ta	Nb, Mo	2–39	No
[18] Benaben et al.	Co	Ca, Fe, Ni, Cu, Zn, As, Se, Mo		No
[19] Krivan	Nb	Ti, V, Cr, Fe, Zr, Mo, Ta, W, Hf	0.3–400	No
[20] Debrun et al.	Rh	Ca, Ti, Cr, Fe, Cu, Zn, Br, Ru, Cd, Sn, Sb, Os, Ir, Pt	0.4–840	No
[21] Nordmann et al.	Na	S		
[22] Krivan	Co	Cr, Cu, Ni	4–600	No
[23] Krivan	Co	Mo	90	No
[24] Krivan	Co	Ti, V, Cr, Fe, Ni, Cu, Zn, Zr, Nb, Mo, Ru, Pd, Sn, W	0.05–570	No
[1] Debrun et al.	Al	Ti, Cr, Ni, Zn	0.01–0.9	No
	Ag	Ti, V, Fe, Cu, As	0.04–50	No
	Au	Ca, Ti, V, Cr, Fe, Ni, Cu, Zn, As	0.04–2100	No
	Co	Fe, Ni, Cu, Zn, Se	2–600	No
	Dy	Ti, V, Cr, Fe, As, Y	0.3–210	No
	Ho	Ca, Ti, Fe, Y, Zr	0.6–320	No
	Ir	Ti, Cr, Fe, Cu, As	0.3–700	No
	Nb	Mo	72	No
	Pr	Ti, Cr, Fe, Ni, As, Y	4–400	No
	Rh	Ca, Ti, Cr, Fe, Cu, Zn, Br	0.5–80	No
	Si	All upper limits		No
	Ta	Nb, Mo	39–2	No
	Tb	Ca, Ti, Cr, Fe, Ni, Y, Mo	1–100	No
[2] Barrandon et al.	Al	All upper limits		No
	Ag	Pd, Cd, Sb, Te, Pb	0.1–17	No
	Au	Ru, Ag, Sn, Sb, Pb	0.26–6.2%	No
	Co	All upper limits		No
	Dy	All upper limits		No
	Ho	All upper limits		No
	Pr	All upper limits		No
	Tb	All upper limits		No
	Ir	Ru, Pd, Sb	50–130	No
	Nb	W	72	No
	Rh	Ru, Cd, Sn, Sb, Ir, Pt	1–360	No

Continued next page

Table 5.3 (*contd.*)

Reference	Matrix analysed	Elements detected	Concentration range, $\mu g/g$	Chemical separation
[2] Barrandon *et al.*	Si	All upper limits		No
	Ta	All upper limits		No
[25] Shibata *et al.*	Al	Fe, V	20–500	No
[26] Benaben *et al.*	Cu	Pd, Pt	1–10	Yes
[27] Faix *et al.*	Nb	Cr, Fe, Cu	0.3–35	Yes
[28] Faix and Krivan	Nb	Cr, Ni, Cu, Zn, Cd	0.1–70	Yes
[29] Sastri and Krivan	Ta, W	As	1–2	Yes
[30] Sastri *et al.*	W	Ti, V, Cr, Fe, As, Zr, Nb	0.03–12	Yes
[31] Lacroix *et al.*	InP	Ti, Cr, Fe, Cu, Zn Ga, Ge, As, Y, Zr, Hg, Pb	0.8–2000 at/cm^3	No
[32] Goethals *et al.*	Fe	Zr	200–1200	No
[33] Dams *et al.*	Fe	Pb	280	Yes
[34] De Brucker *et al.*	Zn	Fe, Cd, Pb	0.2–50	Yes
[11] De Brucker	Zn	Tl	0.7–3.3	Yes
[35] Vandecasteele *et al.*	Cu, Ni, Al	S	<0.5–3	Yes
[36] Schmid *et al.*	Ta	Si	2–9	Yes

5.1.4 Instrumental proton activation analysis

Charged particle activation analysis is most useful when the matrix considered does not yield a high level of induced radioactivity. A purely instrumental analysis is then feasible by γ-ray spectrometry with a germanium semiconductor detector. Whether irradiation of a metal sample will yield a high level of induced radioactivity can easily be predicted. First, with the aid of the Chart of the Nuclides the (p,n), (p,2n), ... (p,α) ... reactions leading to γ-ray emitting radionuclides with a half-life in excess of a few minutes are written down and the thresholds of these reactions are calculated. If the thresholds are in excess of 10 MeV, the incident energy can be chosen below the thresholds of the nuclear reactions of the matrix, in order to allow an instrumental analysis with sufficiently low detection limits for a number of elements. Tables 5.1. and 5.3 may also be useful in deciding whether an instrumental analysis is feasible.

5.1.4.1 *Multi-element analysis of rhodium [20]*

Rhodium is very important in industry, in both pure and alloyed form, the purity needed varying with the type of industrial application. For instance, rhodium used for making thermocouples must be as pure as possible, because the Rh/Pt alloy is fragile when impure rhodium is used. The control of the purity of rhodium is a difficult problem. Methods that require sample dissolution are difficult to apply since rhodium is hard to dissolve. Moreover, thermal neutron activation of rhodium is difficult, because:

— the cross-section for thermal neutrons is high, so errors due to neutron attenuation occur; these systematic errors must be corrected for, or must be avoided by analysis small samples;
— iridium, which is a major impurity of rhodium, becomes highly radioactive upon

irradiation with thermal neutrons; the long-lived ^{192}Ir results in poor detection limits for many elements.

Debrun *et al.* [20] showed that when rhodium is irradiated with 10 MeV protons, the induced γ-radioactivity is low, so a purely instrumental analysis by direct γ-ray spectrometry is possible. Indeed, only ^{103}Pd ($t_{1/2}=17.0$ d), ^{102}Rh ($t_{1/2}=206.0$ d) and ^{102m}Rh ($t_{1/2}=2.89$ y) are produced by the nuclear reactions:

$$^{103}Rh(p,n)^{103}Pd \qquad E_T = 1.35\,MeV$$
$$^{103}Rh(p,d)^{102,\,102m}Rh \qquad E_T = 7.17\,MeV$$
$$^{103}Rh(p,pn)^{102,\,102m}Rh \qquad E_T = 9.42\,MeV$$

Debrun *et al.* [20] determined experimentally the excitation functions for the production of ^{103}Pd and ^{102}Rh, no ^{102m}Rh being detected. The ^{103}Rh(p,pn)^{102}Rh + ^{103}Rh(p,d)^{102}Rh reactions have an experimental threshold at about 10 MeV. The ^{103}Rh(p,n)^{103}Pd reaction has an excitation function with a practical threshold around 2 MeV and a maximum at 10 MeV, so the production of ^{103}Pd cannot be avoided. ^{103}Pd emits γ-rays of 295.2, 357.6 and 497.2 keV with abundances of 0.005, 0.037 and 0.0062%, respectively. Because of these low abundances the activity, expressed in γ-rays emitted per min, of a rhodium sample irradiated with 10 MeV protons, is very low, as shown in Table 5.1. Consequently, the elements that yield radionuclides emitting γ-rays with an energy above 497 keV can be determined with almost the same low detection limits as if there were no interferences. For elements such as vanadium, arsenic, rubidium, iodine, gold and thallium, for which the most intense γ-rays are below 497 keV, the detection limits will be worse because of the Compton continuum due to ^{103}Pd.

The samples, placed behind a Havar monitor foil and behind an aluminium foil, were irradiated with 11 MeV protons. The metal foils reduce the energy to 10 MeV. The ^{56}Fe(p,n)^{56}Co reaction was used for beam intensity monitoring. The irradiations lasted for 1 hr at 0.6–2 μA intensity. After etching in concentrated sulphuric acid the samples were measured several times with a Ge(Li) γ-ray detector.

Table 5.4 shows the experimental detection limits. In spite of the presence of ^{103}Pd and of several impurities the detection limits are almost all below 10 μg/g and most are below 1 μg/g. Table 5.5 gives experimental results for 3 different rhodium samples. The 95% confidence limits given were deduced from the results obtained by making use of the different radionuclides produced from the same element, of the different γ-rays for the same radionuclide, and finally of different measurements. These results indicate good precision and accuracy for chromium, iron, copper, zinc, bromine, ruthenium and platinum. The precision and/or accuracy are somewhat worse for cadmium, tin, antimony (which are close to the detection limit) and iridium, and mainly for calcium and titanium.

5.1.4.2 *Determination of zirconium in nodular cast iron [32]*

During the production of nodular cast iron, i.e. grey cast iron in which the graphite occurs as nodules, zirconium is sometimes added to the melt in order to obtain a

Table 5.4 — Experimental detection limits (μg/g) for a rhodium sample*. (Reprinted with permission, from J. L. Debrun, J. N. Barrandon, P. Benaben and Ch. Rouxel, *Anal. Chem.*, 1975, **47**, 637. Copyright 1975 American Chemical Society)

Element	Radionuclide	Detection limit	Element	Radionuclide	Detection limit
Ca	^{44}Sc	0.15	Ru	^{99m}Rh	0.3
Ti	^{48}V	0.06	Pd	^{105}Ag	1.7
V	^{51}Cr	0.5	Ag	^{107}Cd	5
Cr	^{52}Mn	0.2	Cd	^{111}In	0.5
Fe	^{56}Co	0.25	Sn	^{122}Sb	1
Ni†	^{55}Co	5	Sb	^{121}Te	0.4
Cu	^{65}Zn	2.5	Te	^{130}I	0.2
Zn	^{66}Ga	0.09	I	^{127}Xe	0.5
Ga	^{69}Ge	0.07	Ba	^{135}La	72
Ge	^{72}As	0.08	W	^{184}Re	11
As	^{75}Se	0.8	Re	^{185}Os	3
Se	^{82}Br	0.2	Ir	^{191}Pt	8
Br	^{79}Kr	0.2	Pt	^{194}Au	3
Rb	^{87m}Sr	0.2	Au	^{197m}Hg	6
Sr	^{86}Y	0.9	Hg	^{198m}Tl, ^{200}Tl	2–7
Zr	^{90}Nb	0.1	Tl	^{203}Pb	3
Nb	^{93m}Mo	0.9	Pb	^{206}Bi	5
Mo	^{95}Tc	0.15			

*Irradiation: 1 hr, 2 μA, 10 MeV protons.
†Nuclear reaction used: ^{58}Ni(p, α)^{55}Co.

Table 5.5 — Analytical results for 3 different industrial rhodium samples. Results are 95% confidence limits, and are expressed in μg/g. (Reprinted with permission, from J. L. Debrun, J. N. Barrandon, P. Benaben and Ch. Rouxel, *Anal. Chem.*, 1975, **47**, 637. Copyright 1975 American Chemical Society)

Element	Radionuclides used	Sample 1	Sample 2	Sample 3
Ca	^{44}Sc, ^{48}Sc	9.4±2.7	21±5	5.2±1.5
Ti	^{48}V	0.5±0.2	0.40±0.15	0.35±0.15
Cr	^{52}Mn	50.0±4.5	16.5±1.5	2.20±0.25
Fe	^{56}Co	81±4	78.0±3.5	32.0±1.2
Cu	^{65}Zn	7.8±0.5	4.3±0.3	8.2±0.5
Zn	^{66}Ga	1.3±0.1	1.90±0.15	2.2±0.2
Br	^{79}Kr	7.10±0.35	50.0±1.8	14.0±0.7
Ru	^{99}Rh, ^{99m}Rh, ^{100}Rh	63±3	20.0±1.2	15.0±1.2
Cd	^{111}In	2.2±0.3	<0.35	<0.35
Sn	^{122}Sb	1.9±0.1	<1	3.0±0.4
Sb	^{121}Te	1.3±0.6	3.6±0.5	2.2±0.4
Ir	^{191}Pt	360±74	840±120	190±35
Pt	^{194}Au, ^{196}Au	340±23	400±30	535±45

maximum number of nodules and a minimum amount of carbides. Zirconium can be determined in cast iron by ICP–AES, but this method requires complete dissolution of the samples, which, in the case of cast iron, is difficult and time-consuming.

Quantitative analysis by X-ray fluorescence (XRF) spectrometry is much more convenient, but requires the use of standards with similar composition and physical properties to the sample. Goethals *et al.* [32] used proton activation analysis to determine zirconium in cast iron samples which were used as standards to establish a calibration graph for XRF.

Table 5.6 summarizes some nuclear data on the nuclear reaction used for the

Table 5.6—Nuclear reaction for the determination of zirconium and nuclear reactions with iron and some of its impurities

Nuclear reaction	E_T, MeV	$t_{1/2}$	Most important γ-rays, keV
$^{90}Zr(p,n)^{90}Nb$	7.0	14.6 hr	141; 1129; 2319
$^{56}Fe(p,n)^{56}Co$	5.5	77.3 d	847; 1038; 1238; 1771
$^{52}Cr(p,n)^{52}Mn$	5.6	5.6 d	744; 936; 1434
$^{94}Mo(p,\alpha n)^{90}Nb$	9.0	14.6 hr	141; 1129; 2319

determination of zirconium and on the nuclear reactions with iron and some of the impurities in the cast iron.

As standards, discs of pure zirconium were used, with the same dimensions (15 mm diameter, 1 mm thick) as the cast iron samples.

During the irradiation the samples and the standards were placed behind a copper beam-intensity monitor foil. The $^{65}Cu(p,n)^{65}Zn$ reaction was used for beam intensity monitoring. The samples and the standards were irradiated with 12 MeV protons, the monitor foil reducing the energy to 11.6 MeV. The irradiations lasted for 30 and 5 min, at beam intensities of 1.0 and 0.2 μA, for the samples and the standards, respectively.

The samples and the standards were measured ca. 3 hr after the irradiation, with a Ge(Li) γ-ray spectrometer, with 4 cm of lead interposed between the samples and the detector to attenuate the γ-rays from ^{56}Co. The 2319 keV peak of ^{90}Nb was used for the calculation of the zirconium concentration.

It was shown that the interference factor for the interference of molybdenum through the $^{94}Mo(p,\alpha n)^{90}Nb$ reaction is less than $2.9 \times 10^{-3}\%$ for the experimental conditions used.

Table 5.7 gives the results for 4 types of cast iron. The relative standard deviation

Table 5.7 — Results for zirconium in nodular cast iron. Concentrations in μg/g

Cast iron	Proton activation			ICP–AES	
	$\bar{x}$	s	n	$\bar{x}$	s
H36	1197	34	4	1220	33
H35	693	33	4	670	16
H34	643	25	4		
H33	212	4	3	195	4

of the individual results, expected from the counting statistics, ranges from 2 to 9%, whereas the experimental relative standard deviation ranges from 2 to 5%. The results of proton activation analysis are in good agreement with those obtained by ICP–AES for the same samples. Moreover, when, in order to obtain a calibration graph for XRF, the Zr K_α-intensity is plotted as a function of the zirconium concentration, a straight line is obtained.

5.1.4.3　Determination of trace elements in cobalt [24]

Cobalt is difficult to analyse by reactor neutron activation analysis, because of the high matrix activity due to the high thermal neutron cross-section and resonance integral for the $^{59}Co(n,\gamma)^{60}Co$ reaction, and because of the long half-life of ^{60}Co.

Table 5.8 gives nuclear reactions that can be induced in cobalt irradiated with

Table 5.8 — Nuclear reactions occurring when cobalt is irradiated with protons. (Reprinted with permission, from V. Krivan, *Talanta*, 1976, **23**, 621. Copyright 1976, Pergamon Press, Oxford.)

Nuclear reaction	E_T, MeV	$t_{1/2}$	γ-Rays, MeV	Intensity, %
$^{59}Co(p,n)^{59}Ni$	1.9	7.5×10^4 y	—	—
$^{59}Co(p,pn)^{58m}Co$	10.7	8.94 hr	0.025	0.03
$^{59}Co(p,pn)^{58}Co$	10.7	70.78 d	0.511	30.0
			0.810	99.4
			0.864	0.7
			1.675	0.5
$^{59}Co(p,p2n)^{57}Co$	19.3	270.0 d	0.014	9.5
			0.122	85.6
			0.136	10.6
			0.570	0.01
			0.692	0.15
$^{59}Co(p,\alpha n)^{55}Fe$	8.1	2.7 y	—	—

protons of energy up to 20 MeV. Instrumental analysis by γ-ray detection is only possible if the production of ^{57}Co, and mainly of ^{58}Co, is avoided. The activation curve of the $^{59}Co(p,pn)^{58}Co$ reaction is given in Fig. 5.2. A proton energy of 12 MeV can be used even for a long irradiation time; for short irradiations up to 13 MeV protons can be used. Table 5.9 gives nuclear data for the nuclear reactions used and the indicator nuclides and Table 5.10 gives possible interfering reactions. At the incident energies used, no nuclear interference occurs in the determination of titanium and chromium. In addition, it was experimentally shown [16, 19, 22] that in the determination of zirconium, niobium, vanadium, iron, molybdenum, tungsten, nickel and copper, no significant interferences occur, even if the interfering element is present in higher concentration. Finally, a detailed analysis of the experimental results obtained with irradiated cobalt samples showed that no detectable interferences occurred in the determination of zinc, ruthenium, palladium, and tin, which could be interfered with by (p,α) or $(p,\alpha n)$ reactions of germanium, palladium, cadmium and tellurium, respectively. The presence of these interfering elements and the extent of the interferences were checked through the corresponding (p,n)

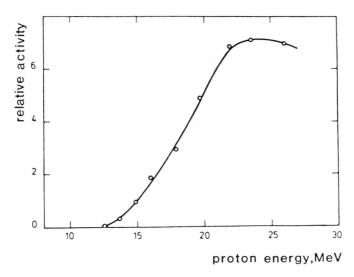

Fig. 5.2 — Activation curve for the $^{59}Co(p,pn)^{58}Co$ reaction. (Reprinted with permission, from V. Krivan, *Talanta*, 1976, **23**, 621. Copyright 1976, Pergamon Press, Oxford.)

reactions, which would yield activities one or two orders of magnitude higher than the interfering reactions. Krivan [24] also compiled possible instrumental interferences. These could be avoided by choosing another γ-ray or counting the sample after the decay of the interfering radionuclide. Table 5.11 gives some results, based on 3 determinations. The samples were irradiated for 5–120 min with 12 or 13 MeV protons at 3.5 μA intensity. For niobium and ruthenium, the precision of the results is rather poor, as the concentrations are close to the detection limit; for the other elements the precision is satisfactory.

5.1.5   Radiochemical proton activation analysis

When the matrix yields a high level of induced radioactivity on irradiation with e.g. 10 MeV protons, an instrumental analysis is not feasible. To allow the determination of trace elements the indicator radionuclides must then be separated from the matrix activities. The radiochemical separations are often done by ion-exchange or liquid–liquid extraction. Sometimes very high decontamination factors are required, since the activity from the matrix is much higher than the activity of the indicator radionuclide.

5.1.5.1   *Simultaneous determination of iron, cadmium, and lead in zinc [34]*
A BCR (Community Bureau of Reference, Commission of the European Communities) programme aimed at the certification of several impurities in zinc metal. Routine methods used for the determination of trace elements in zinc, such as atomic-absorption spectrometry, inductively-coupled plasma atomic-emission spec-

Table 5.9 — Data on the nuclear reactions used and indicator radionuclides. (Reprinted with permission, from V. Krivan, *Talanta*, 1976, **23**, 621. Copyright 1976, Pergamon Press, Oxford)

Element determined	Principal reactions	Isotopic abundance, %	E_T, MeV	Half-life	Major γ-rays, MeV	Intensity, %
Ti	$^{48}Ti(p,n)^{48}V$	73.7	4.9	15.97 d	0.9443	8.0
					0.9835	100.0
					1.3116	98.0
V	$^{51}V(p,n)^{51}Cr$	99.75	1.5	27.7 d	0.3201	9.8
Cr	$^{52}Cr(p,n)^{52m}Mn$	83.79	5.7	21.0 min	1.4343	100.0
	$^{52}Cr(p,n)^{52}Mn$	83.79	5.7	5.7 d	0.7442	85.0
					0.9356	93.0
					1.4343	100.0
Fe	$^{56}Fe(p,n)^{56}Co$	91.7	5.5	77.3 d	0.8467	99.99
					1.0378	14.0
					1.2383	67.6
					1.7715	15.7
					2.5986	16.9
Ni	$^{60}Ni(p,n)^{60}Cu$	26.42	7.0	23.0 min	0.8260	19.2
					1.3325	87.3
					1.7920	44.9
Cu	$^{63}Cu(p,n)^{63}Zn$	69.1	4.2	38.4 min	0.6696	8.5
					0.9619	6.7
Zn	$^{66}Zn(p,n)^{66}Ga$	27.8	6.1	9.3 hr	1.0393	37.3
					2.7523	22.8
Zr	$^{90}Zr(p,n)^{90}Nb$	51.4	7.0	14.6 hr	0.1412	67.0
					1.1291	92.0
					2.1860	18.0
					2.3190	82.0
Nb	$^{93}Nb(p,n)^{93m}Mo$	100.0	1.2	6.9 hr	0.2632	61.2
					0.6846	91.9
					1.4772	99.4
Mo	$^{94}Mo(p,n)^{94m}Tc$	9.1	5.0	53.0 min	0.8709	91.0
					1.5220	5.4
					1.8688	5.3
	$^{94}Mo(p,n)^{94}Tc$	9.1	5.0	4.9 hr	0.7026	100.0
					0.8497	100.0
					0.8709	100.0
					0.9162	6.8
	$^{95}Mo(p,n)^{95}Tc$	15.9	2.5	20.0 hr	0.7658	94.0
	$^{96}Mo(p,n)^{96}Tc$	16.7	3.7	4.3 d	0.7783	100.0
					0.8128	82.0
					0.8503	99.0
					1.1272	15.0
Ru	$^{100}Ru(p,n)^{100}Rh$	12.6	4.4	20.0 hr	0.4462	11.5
					0.5396	80.0
					0.8225	20.6
					1.1071	13.5
					1.3621	15.3
					1.5534	21.0
					1.9297	12.4
					2.3761	35.7
	$^{101}Ru(p,n)^{101m}Rh$	17.1	1.3	4.4 d	0.3068	94.0
Pd	$^{104}Pd(p,n)^{104m}Ag$	11.0	4.9	33.5 min	0.5558	91.0
	$^{104}Pd(p,n)^{104}Ag$	11.0	4.9	69.2 min	0.5558	92.5
					0.7674	66.3
					0.9232	16.3
					0.9259	12.5
					0.9416	25.0
					1.5261	6.2

Continued next page

Table 5.9 (*contd.*)

Element determined	Principal reactions	Isotopic abundance, %	E_T, MeV	Half-life	Major γ-rays, MeV	Intensity, %
Sn	$^{116}Sn(p,n)^{116}Sb$	14.4	5.3	16.0 min	0.9330	23.0
					1.2937	88.0
					2.2298	11.5
	$^{120}Sn(p,n)^{120m}Sb$	32.8	3.5	5.76 d	0.0898	77.0
					0.1972	89.0
					1.0230	99.0
					1.1713	100.0
W	$^{182}W(p,n)^{182m}Re$	26.3	3.6	13.0 hr	0.1001	17.4
					1.1213	38.5
					1.1890	18.4
					1.2214	30.5
	$^{182}W(p,n)^{182}Re$	26.3	3.6	64.0 hr	0.1001	15.0
					0.1489	15.0
					0.1784	18.0
					0.2292	25.0
					0.2564	13.0
					1.1220	20.0
					1.1895	7.9
					1.2218	15.0

Table 5.10 — Nuclear interferences. (Reprinted with permission, from V. Krivan, *Talanta*, 1976, **23**, 621. Copyright 1976, Pergamon Press, Oxford)

Element determined	Activation reaction used	Interfering reaction	E_T MeV	Isotopic abundance, %
Ti	$^{48}Ti(p,n)^{48}V$	$^{52}Cr(p,\alpha n)^{48}V$	14.4	83.79
V	$^{51}V(p,n)^{51}Cr$	$^{55}Mn(p,\alpha n)^{51}Cr$	9.2	100.00
		$^{52}Cr(p,pn)^{51}Cr$	12.2	83.79
Cr	$^{52}Cr(p,n)^{52}Mn$	$^{56}Fe(p,\alpha n)^{52}Mn$	13.3	91.7
Fe	$^{56}Fe(p,n)^{56}Co$	$^{60}Ni(p,\alpha n)^{56}Co$	11.8	26.23
Ni	$^{60}Ni(p,n)^{60}Cu$	$^{64}Zn(p,\alpha n)^{60}Cu$	11.1	48.9
Cu	$^{63}Cu(p,n)^{63}Zn$	$^{64}Zn(p,pn)^{63}Zn$	12.0	48.9
Zn	$^{66}Zn(p,n)^{66}Ga$	$^{70}Ge(p,\alpha n)^{66}Ga$	10.1	20.7
Zr	$^{90}Zr(p,n)^{90}Nb$	$^{94}Mo(p,\alpha n)^{95}Nb$	9.1	9.1
Nb	$^{93}Nb(p,n)^{93m}Mo$	$^{94}Mo(p,pn)^{93m}Mo$	9.8	9.1
Mo	$^{94}Mo(p,n)^{94}Tc$	$^{98}Ru(p,\alpha n)^{94}Tc$	7.4	1.9
	$^{95}Mo(p,n)^{95}Tc$	$^{96}Ru(p,2p)^{95}Tc$	7.5	5.5
		$^{98}Ru(p,\alpha)^{95}Tc$	$Q>0$	1.9
		$^{100}Ru(p,\alpha n)^{96}Tc$	7.4	12.7
	$^{96}Mo(p,n)^{96}Tc$	$^{99}Ru(p,\alpha)^{96}Tc$	$Q>0$	12.7
		$^{100}Ru(p,\alpha n)^{96}Tc$	6.7	12.6
Ru	$^{100}Ru(p,n)^{100}Rh$	$^{104}Pd(p,\alpha n)^{100}Rh$	7.1	11.0
	$^{101}Ru(p,n)^{101m}Rh$	$^{104}Pd(p,\alpha)^{101m}Rh$	$Q>0$	11.0
		$^{105}Pd(p,\alpha n)^{101m}Rh$	4.2	22.2
Pd	$^{104}Pd(p,n)^{104}Ag$	$^{108}Cd(p,\alpha n)^{104}Ag$	7.2	0.9
	$^{106}Pd(p,n)^{106m}Ag$	$^{107}Ag(p,pn)^{106m}Ag$	9.6	51.83
		$^{110}Cd(p,\alpha n)^{106m}Ag$	6.7	12.4
Sn	$^{116}Sn(p,n)^{116}Sb$	$^{120}Te(p,\alpha n)^{116}Sb$	5.6	0.09
	$^{120}Sn(p,n)^{120m}Sb$	$^{121}Sb(p,pn)^{120m}Sb$	9.3	57.3
		$^{123}Te(p,\alpha)^{120m}Te$	$Q>0$	0.87
		$^{124}Te(p,\alpha n)^{120m}Te$	5.3	4.6
W	$^{182}W(p,n)^{182}Re$	$^{186}Os(p,\alpha n)^{182}Re$	0.8	1.6
		$^{187}Os(p,\alpha 2n)^{182}Re$	7.1	1.6

Table 5.11 — Concentrations of trace elements in cobalt metal. (Reprinted with permission, from V. Krivan, *Talanta*, 1976, **23**, 621.Copyright 1976, Pergamon Press, Oxford.)

Trace element	Indicator radionuclide	Counted γ-rays, keV	Concentration found by proton activation, $\mu g/g$	Detection limit, $\mu g/g$
Ti	^{48}V	983	0.6±0.1	0.04
		1312		
V	^{51}Cr	320	<1.4	1.4
Cr	^{52m}Mn	1434	4.6±0.6	0.4
	^{52}Mn	1434	4.7±0.5	0.15
Fe	^{56}Co	1238	15.4±2.3	0.3
		1771		
		2598		
Ni	^{60}Cu	826	568±61	3.9
		1332		
		1792		
Cu	^{66}Zn	670	38.2±5.8	12.8
Zn	^{65}Ga	2752	0.4±0.1	0.05
Zr	^{90}Nb	2319	<0.05	0.05
Nb	^{93m}Mo	1477	0.9±0.7	0.4
Mo	^{94}Tc	703	87.0±4.2	1.1
		871		
	^{95}Tc	766	93.4±3.8	0.09
	^{96}Tc	778	89.1±3.9	0.1
Ru	^{100}Rh	1553	0.4±0.3	0.15
Pd	^{104,104m}Ag	556	<0.6	0.6
Sn	^{120m}Sb	1023	<0.6	0.6
W	^{182}Re	1121	23.7±3.7	0.8
		1189		
		1221		

trometry, and spectrophotometry, all require dissolution of the sample, with its attendant risks of contamination and high blanks. Therefore, for the materials with the lowest trace element concentration, for which there was a lack of accurate procedures, proton activation analysis was applied.

The principal nuclear reactions of the elements of interest and the interfering nuclear reactions for activation with 18 MeV protons are given in Table 5.12. An incident energy of 18 MeV was chosen in order to obtain sufficiently low detection limits, mainly for lead. The determination of iron and lead is free from nuclear interferences. Nuclear interference from tin must be considered in determination of cadmium: the interference factor is 0.6%. Table 5.13 gives the nuclear reactions which are energetically possible with 18 MeV protons and a zinc matrix. The high matrix activity makes a purely instrumental analysis impossible, so a post-irradiation chemical separation of ^{56}Co, ^{111}In and ^{206}Bi from radioisotopes of gallium and zinc is required.

The samples, placed behind a nickel foil serving as a beam intensity monitor, were irradiated for 1 hr with 18 MeV protons at 1 μA intensity. Iron(III) oxide, cadmium(II) oxide, and lead(II) oxide were used as standards. After irradiation, in order to remove possible surface contamination, the samples were chemically etched in 6M nitric acid.

Table 5.12 — Principal nuclear reactions and nuclear data for the elements determined. (Reprinted by permission, from N. De Brucker, K. Strijckmans and C. Vandecasteele, *Anal. Chim. Acta*, 1987, **195**, 323. Copyright 1987, Elsevier Science Publishers.)

Nuclear reaction	E_T, MeV	Half-life, days	γ-Rays used, keV	Interfering reactions	E_T, MeV
$^{56}Fe(p,n)^{56}Co$	5	77.3	846.8	None	
$^{57}Fe(p,2n)^{56}Co$	13				
$^{111}Cd(p,n)^{111}In$	2	2.8	171.3; 245.4	$^{114}Sn(p,\alpha)^{111}In$	$Q>0$
$^{112}Cd(p,2n)^{111}In$	11			$^{115}Sn(p,\alpha n)^{111}In$	5
				$^{116}Sn(p,\alpha 2n)^{111}In$	14
$^{206}Pb(p,n)^{206}Bi$	4	6.2	803.1; 881.0	None	
$^{207}Pb(p,2n)^{206}Bi$	11				

Table 5.13 — Nuclear reactions of zinc irradiated with 18 MeV protons. (Reprinted by permission, from N. De Brucker, K. Strijckmans and C. Vandecasteele, *Anal. Chim. Acta*, 1987, **195**, 323. Copyright 1987, Elsevier Science Publishers.)

Nuclear reaction	E_T, MeV	Half-life, days	Major γ-rays, keV
$^{66}Zn(p,n)^{66}Ga$	6	0.39	1039.4; 833.7
$^{67}Zn(p,2n)^{66}Ga$	13		
$^{67}Zn(p,n)^{67}Ga$	2	3.3	93.3; 184.6
$^{68}Zn(p,2n)^{67}Ga$	12		300.2; 393.6
$^{66}Zn(p,pn)^{65}Zn$	11	243.8	1115.5

After chemical etching, the zinc samples were dissolved in concentrated hydrochloric acid, together with inactive carrier. This solution was transferred to a cation-exchange column and eluted with 8M hydrochloric acid. Gallium was retained, whereas cobalt, indium, bismuth and zinc were eluted. The eluate was then divided into two equal fractions (A and B). Fraction A was used for the separation of cobalt from zinc by anion-exchange in 3M hydrochloric acid and fraction B for the separation of indium and bismuth from zinc by precipitation with thionalide. For details of the experimental procedure reference may be made to De Brucker *et al.* [34].

The detection limits of the method are 0.03, 0.008 and 0.1 μg/g for iron, cadmium and lead, respectively. These detection limits relate to a 1 hr irradiation with a 1 μA beam of 18 MeV protons, followed by a 15 hr measurement after a decay time of 1 day.

Table 5.14 gives the results for the BCR reference material 321, unalloyed zinc, together with the provisional certification values and the number of laboratories that provided the accepted results. The results of proton activation analysis are in good agreement with the provisional certified values and their precision ranges from 1 to 7%. At these rather low concentration levels, very few methods could meet the requirements for certification.

Table 5.14 — Iron, cadmium and lead in zinc (BCR Reference Material 321, Unalloyed Zinc) (Reprinted by permission, from N. De Brucker, K. Strijckmans and C. Vandecasteele, *Anal. Chim. Acta*, 1987, **195**, 323. Copyright 1987, Elsevier Science Publishers.)

Metal	Concentration found, μg/g		Provisional certification $\pm$uncertainty	Number of laboratories
	$\bar{x}$	s^*		
Fe	2.23	0.14	2.3 $\pm$0.2	11
Cd	0.215	0.015	0.22$\pm$0.04	5
Pd	4.641	0.061	4.7 $\pm$0.2	6

5.1.5.2 Determination of lead in cast iron

Dams *et al.* [33] determined lead in cast iron by radiochemical proton activation analysis with the ^{205,206,207}Pb(p,xn)^{206}Bi reactions.

Discs of cast iron were irradiated for 10–20 min with a 1 μA beam of 23 MeV protons. After the irradiation the samples were dissolved in a mixture of nitric and perchloric acids to which bismuth carrier was added. The bismuth was extracted with antimony diethyldithiocarbamate in chloroform. After back-extraction with hydrochloric acid the ^{206}Bi was measured by γ-spectrometry. Discs of pure lead were used as the standard.

Ten analyses yielded 283.6 μg/g with a standard deviation of 5.3 μg/g, in good agreement with the results obtained by ETA–AAS after dissolution of the samples.

5.1.5.3 Determination of titanium, vanadium, chromium, iron, arsenic, zirconium and niobium in tungsten [30]

Tungsten, because of its physical and chemical properties, is widely used in different fields of science and technology, most applications being based on the high melting point, low vapour pressure, and high resistance to corrosion. Since its properties depend on the purity, methods for trace analytical characterization of tungsten are required. Methods in common use (spectrophotometry, AAS, and AES) allow detection limits at the 1 μg/g or sometimes at the 0.1 μg/g level in tungsten. In neutron activation analysis a very high matrix activity is produced owing to the high cross-section of the ^{186}W(n,γ)^{187}W reaction. ^{187}W has a 23.8 hr half-life, so radiochemical separations can only be performed after a decay period of several days, or under special conditions ensuring radiation protection. Sastri *et al.* [30] showed that radiochemical proton activation analysis is a useful alternative to neutron activation analysis in this case.

Table 5.15 gives the nuclear reactions that are energetically possible when tungsten is irradiated with 13 MeV protons, and Table 5.16 gives the principal reactions of the elements of interest. It appears from Table 5.15 that irradiation of tungsten with protons yields several radionuclides of rhenium, causing high matrix activity. An instrumental analysis is thus impossible, and a radiochemical separation is required. Sastri *et al.* [30] used anion-exchange in hydrofluoric acid medium.

Table 5.15 — Nuclear reactions of tungsten. (Reproduced from C. S. Sastri, R. Caletka and V. Krivan, *J. Trace Microprobe Tech.*, 1983, **1**, 293, courtesy of Marcel Dekker, Inc.)

Nuclear reaction	E_T, MeV	Half-life
$^{180}W(p,n)^{180}Re$	4.6	2.43 min
$^{182}W(p,n)^{182m}Re$	3.6	13 hr
$^{182}W(p,n)^{182}Re$	3.6	2.67 d
$^{182}W(p,2n)^{181}Re$	11.3	20 hr
$^{183}W(p,n)^{182}Re$	1.3	71 d
$^{184}W(p,n)^{184m}Re$	3.6	165 d
$^{184}W(p,n)^{184}Re$	3.6	38 d

Table 5.16 — Nuclear reactions of the elements of interest. (Reproduced from C. S. Sastri, R. Caletka and V. Krivan, *J. Trace Microprobe Tech.*, 1983, **1**, 293, courtesy of Marcel Dekker, Inc.)

Element	Nuclear reactions	E_T, MeV	Half-life	γ-Ray energy used, keV
Ti	$^{48}Ti(p,n)^{48}V$	4.9	15.97 d	1311
V	$^{51}V(p,n)^{51}Cr$	1.5	27.7 d	320
Cr	$^{52}Cr(p,n)^{52}Mn$	5.6	5.6 d	744
Fe	$^{56}Fe(p,n)^{56}Co$	5.5	78.7 d	846
As	$^{75}As(p,n)^{75}Se$	1.6	120 d	264
Zr	$^{92}Zr(p,n)^{92m}Nb$	2.8	10.16 d	934
Nb	$^{93}Nb(p,n)^{93m}Mo$	1.2	6.95 hr	684

Titanium, vanadium, chromium, iron, zirconium and niobium as the metals, and arsenic trioxide mixed with graphite, were used as standards.

The samples were irradiated for 15–60 min with 13 MeV protons at 5 μA beam intensity, and the standards for 1–2 min at 0.1–0.2 μA.

After irradiation, in order to remove surface contamination the samples were etched in a hydrofluoric acid–nitric acid mixture, and they were also dissolved in a hydrofluoric acid–nitric acid mixture. Carriers of vanadium, chromium, manganese, cobalt, niobium and molybdenum were added, and the solution was evaporated almost to dryness. The evaporation was repeated after addition of concentrated hydrofluoric acid, and the residue was dissolved in 10M hydrofluoric acid. The chemical separation was done by anion-exchange on a column of Dowex 1–×8, prewashed with 10M hydrofluoric acid. After application of the sample, the column was washed with 10M hydrofluoric acid to elute vanadium, chromium, manganese, cobalt, and selenium. In the next step, the tungsten matrix together with the nuclides of tungsten produced by (n,γ) reactions of the matrix, were eluted with 6M hydrochloric acid/3M hydrofluoric acid. Finally, niobium and molybdenum were eluted with 8M hydrochloric acid/0.2M hydrofluoric acid and with 11M hydrochloric acid, respectively. The activity of the eluates was measured with a Ge(Li) detector.

Although rhenium is almost quantitatively retained on the anion-exchanger, it may sometimes be necessary to purify the molybdenum fraction further from rhenium.

Table 5.17 gives the results obtained for high-purity tungsten, along with the detection limits. Except for niobium the detection limits are in the 20–65 ng/g range.

Table 5.17 — Determination of trace elements in high-purity tungsten. (Reproduced from C. S. Sastri, R. Caletka and V. Krivan, *J. Trace Microprobe Tech.*, 1983, **1**, 293, courtesy of Marcel Dekker, Inc.)

Element	Concentration, μg/g		Detection limit*, ng/g
	$\bar{x}$	s	
Ti	1.42	0.12	25
V	0.17	0.02	20
Cr	0.20	0.04	50
Fe	11.8	0.6	50
As	2.1	0.2	65
Zr	0.03	0.01	30
Nb	1.5	0.4	350

* Irradiation for 2 hr at 2 μA. Measurement for 40000 sec.

5.2 ACTIVATION ANALYSIS WITH LIGHT PROJECTILES OTHER THAN PROTONS

5.2.1 Nuclear reactions, detection limits

Light projectiles other than protons can also be used to determine medium and heavy elements.

Kormali and Schweikert [4] have made a survey of the application of 20 MeV protons and deuterons and of 40 MeV helium-3 and helium-4 for the determination of medium Z elements. It was shown that deuteron-induced reactions provide 3–20 times higher thick-target yields than the proton-induced reactions do. Because of the uniformly low threshold energies, however, selective activation of the element of interest is difficult. The thick-target yields for helium-3 or helium-4 activation are lower than those for proton and deuteron activation. Sastri *et al.* [37] investigated nuclear reactions for the determination of magnesium, aluminium, titanium, vanadium, chromium, manganese, iron, nickel, zinc, zirconium, niobium, molybdenum, and silver by activation analysis with 14 MeV helium-3. It was shown that for an irradiation of 1 hr or 1 half-life, whichever is the shorter, at 2 μA the detection limits range from 3 to 400 ng/g. Table 5.18 gives the most sensitive nuclear reactions, along with the detection limits and the interfering reactions. The 14 MeV irradiation energy corresponds to the Coulomb threshold energy for helium-3 nuclei and a target nucleus with $Z{\sim}42$. Elements with $Z{>}42$ can thus undergo nuclear reactions only because of the tunnelling effect. Activation analysis with helium-3 can hence be used for selective determination of low and medium Z elements ($Z{<}42$) in high Z matrices.

Giovagnoli *et al.* [38] studied nuclear reactions induced by helium-4 with magnesium, aluminium and phosphorus.

5.2.2 Applications

Sastri *et al.* [39, 40] used helium-3 activation to determine vanadium, iron, nickel, zinc and molybdenum in niobium, tantalum and tungsten.

The determination of phosphorus [41] and calcium [42] by helium-4 activation will be described in detail.

Table 5.18 — Nuclear reactions and detection limits for 14 MeV helium-3 activation. (Rerpinted with permission, from C. S. Sastri, H. Petri and G. Erdtmann, *Anal. Chem.*, 1977, **49**, 1510. Copyright 1977, American Chemical Society.)

Element	Nuclear reaction	E_T, MeV	Half-life		Detection limit*, ng/g	Interfering reaction
Mg	$^{26}Mg(^3He,2p)^{27}Mg$	1.4	9.48	min	220	—
Al	$^{27}Al(^3He,2p)^{28}Al$	$Q>0$	2.24	min	17	$^{26}Mg(^3He,p)^{28}Al$
Ti	$^{48}Ti(^3He,2n)^{49}Cr$	4.6	41.9	min	3	$^{50}Cr(^3He,\alpha)^{49}Cr$
V	$^{51}V(^3He,2n)^{52m}Mn$	2.9	21.3	min	22	$^{52}Cr(^3He,p2n)^{52m}Mn$
Cr	$^{52}Cr(^3He,2n)^{53}Fe$	6.0	8.51	min	120	$^{54}Fe(^3He,\alpha)^{53}Fe$
Mn	$^{55}Mn(^3He,2p)^{56}Mn$	0.4	2.576	hr	40	$^{54}Cr(^3He,p)^{56}Mn$
Fe	$^{56}Fe(^3He,p)^{58}Co$	$Q>0$	71.3	d	270	$^{59}Co(^3He,\alpha)^{58}Co$
Ni	$^{60}Ni(^3He,pn)^{61}Cu$	3.0	3.41	hr	40	$^{59}Co(^3He,n)^{61}Cu$
						$^{63}Cu(^4He,\alpha n)^{61}Cu$
Zn	$^{64}Zn(^3He,pn)^{65}Ga$	2.9	15.2	min	10	$^{63}Cu(^3He,n)^{65}Ga$
Zr	$^{96}Zr(^3He,pn)^{97}Nb$	0.2	72.0	min	320	—
Nb	$^{93}Nb(^3He,2n)^{94}Tc$	4.4	4.88	hr	30	$^{92}Mo(^3He,2p)^{94}Tc$
Mo	$^{95}Mo(^3He,n)^{99m}Tc$	1.2	6.02	hr	70	—
Ag	$^{107}Ag(^3He,2n)^{108m}In$	5.7	58	min	440	$^{108}Cd(^3He,p2n)^{108}In$

*For irradiation of 1 hr or 1 half-life, at 2 μA intensity.

5.2.2.1 Determination of phosphorus in aluminium-silicon alloy by helium-4 activation

The $^{31}P(\alpha,n)^{34m}Cl$ reaction is the most suitable nuclear reaction for the determination of phosphorus. Table 5.19 gives some information on this nuclear reaction

Table 5.19 — Nuclear reaction for the determination of phosphorus. The threshold energy in MeV is given in brackets

Reaction		Interfering reactions	
$^{31}P(\alpha,n)^{34m}Cl$	(5.8)	$^{35}Cl(\alpha,\alpha n)^{34m}Cl$	(14.5)
		$^{32}S(\alpha,d)^{34m}Cl$	(13.0)

and the nuclear interferences. The radionuclide produced, ^{34m}Cl, has a 32.2 min half-life and emits positrons and γ-rays of 145.7 (35.8%), 1177.4 (14.2%) and 2128.5 keV (48.4%). Goethals *et al.* [41] showed that for an incident energy of 20 MeV, 290 μg/g of chlorine and 360 μg/g of sulphur yield the same ^{34m}Cl activity as 1 μg/g of phosphorus.

Phosphorus is of economic and technological importance in aluminium–silicon alloy: it regulates the mechanism of solidification of the eutectic (12.5% Si) and nearly eutectic alloy, and it grain-refines the primary silicon in the hypereutectic system (15–25% Si). Goethals *et al.* [41] developed a method for the determination of phosphorus in aluminium–silicon alloy by using the $^{31}P(\alpha,n)^{34m}Cl$ reaction.

Table 5.20 gives the irradiation conditions and information on the post-irradiation chemical etching. As a standard, aluminium powder mixed with disodium hydrogen phosphate was used. The mixture was carefully homogenized and pressed into a pellet. The samples were measured for 30 min, 30 min after the irradiation, by means of a Ge(Li) γ-ray spectrometer. The delay was sufficient to allow decay of ^{30}P ($t_{1/2}=2.5$ min) produced by the $^{27}Al(\alpha,n)^{30}P$ reaction. The standards were measured for 5 min, 90 min after irradiation, in the same geometry as the samples.

Table 5.20 — Irradiation and chemical etching after the irradiation

	Sample	Standard
α-Particle energy (MeV)	25	25
Intensity (μA)	2–3	2–3
Irradiation time (min)	20–30	1
Monitor foil	Cu 22.3 mg/cm^2	Cu 22.3 mg/cm^2
Additional foils		Al 5.4 mg/cm^2
Etch	1/1(v/v)14M HNO$_3$/50%HF 2–5 min, room temp.	
Thickness removed	5.4 mg/cm^2	
Effective incident energy (MeV)	20.2	20.2
Residual range (mg/cm^2)	53	

Table 5.21 gives the results for aluminium–silicon alloys of different origin and with different silicon concentrations. For phosphorus concentrations between 1.7 and 25 μg/g the relative standard deviation ranged from 6 to 17%. The same alloys were analysed by neutron activation analysis using the ^{31}P(n,γ)^{32}P reaction. ^{32}P ($t_{1/2}$ = 14.3 d, $E_{\beta-}$ = 1710 keV) was separated by precipitation as ammonium molybdo-phosphate and measured with a GM-counter. The results are also given in Table 5.21 and are in good agreement with those obtained by α-particle activation analysis. The α-particle activation has the advantage of being instrumental and more rapid than neutron activation.

Table 5.21 — Determination of phosphorus in aluminium–silicon alloy.
Results in μg/g

Alloy	Si, %	α-Activation			Neutron activation		
		$\bar{x}$	s	n	$\bar{x}$	s	n
AS 3	3	1.72	0.11	2	1.785	0.047	2
LMG	9.5	4.88	0.85	10	5.091	0.099	5
AS 13	13	5.849	0.086	2	5.645	0.045	2
VAW	—	8.41	0.76	10	8.28	0.16	5
BCR	12	25.0	1.7	9	23.9	1.8	5

5.2.2.2 Determination of calcium in cast iron by helium-4 activation

Nuclear reactions for the determination of calcium should preferably start from ^{40}Ca (96.94% isotopic abundance). Table 5.22 gives the nuclear reactions of ^{40}Ca that yield radionuclides with a half-life longer than 1 min. The ^{40}Ca(α, p)^{43}Sc reaction has the advantage that the 3.89 hr half-life product allows a post-irradiation radiochemi-cal separation. Table 5.23 lists the nuclear interferences. Owing to the low isotopic abundance of ^{40}K (0.012%), interference by the ^{40}K(α, n)^{43}Sc reaction is negligible. The ^{41}K(α, 2n)^{43}Sc and ^{45}Sc(α, α2n)^{43}Sc reactions do not occur for incident energies below 14.4 MeV.

Table 5.22 — Nuclear reactions of ^{40}Ca. (Reprinted by permission, from C. Vandecasteele, F. Alluyn, J. Dewaele and R. Dams, *Anal. Chem.*, 1985, **57**, 2549. Copyright 1985, American Chemical Society)

Reaction	Threshold energy, MeV	Radionuclide formed	$t_{1/2}$	Radiation emitted
^{40}Ca(^{3}He,p)^{42m}Sc	$Q > 0$	^{42m}Sc	61.6 sec	β^+, γ
^{40}Ca(α,pn)^{42m}Sc	17.3	^{42m}Sc		
^{40}Ca(d,α)^{38}K	$Q > 0$	^{38}K	7.63 min	β^+, γ
^{40}Ca(α,p)^{43}Sc	3.8	^{43}Sc	3.89 hr	β^+, γ (372.8 keV)

Table 5.23 — Nuclear interferences. (Reprinted by permission, from C. Vandecasteele, F. Alluyn, J. Dewaele and R. Dams, *Anal. Chem.*, 1985, **57**, 2549. Copyright 1985, American Chemical Society)

Nuclear reaction	Threshold energy, MeV	Isotopic abundance, %
^{40}Ca(α,p)^{43}Sc	3.8	96.94
^{40}K(α,n)^{43}Sc	3.3	0.012
^{41}K(α,2n)^{43}Sc	14.4	6.7
^{45}Sc(α,α2n)^{43}Sc	22.9	100

During the production of nodular cast iron, certain post-inoculants are added to the melt to obtain a maximum number of nodules and a minimum amount of carbides. Post-inoculation with calcium-bearing ferrosilicon is successful, although the precise role of calcium in the nucleation mechanism is not clear. Calcium may also have a negative effect, even in minute quantities, since it promotes the formation of chunks of graphite in heavy ductile-iron castings. In practice, the amount of calcium to be added should be optimized. Therefore, for a better control of the production process, an accurate determination of calcium in nodular cast iron at the μg/g level is necessary. Vandecasteele *et al.* [42] chose wavelength-dispersive X-ray fluorescence spectrometry (XRF) at the Ca K_α line (3.69 keV), because it is purely instrumental and fast, and sample preparation is relatively easy. Since certified reference materials for calcium in nodular cast iron were not available, the calcium concentration in some cast iron samples was determined by CPAA in order to establish a calibration graph for XRF.

The ^{40}Ca(α,p)^{43}Sc reaction was used to determine calcium by α-particle activation analysis. Calcium carbonate powder was used as a standard. Table 5.24 summarizes the irradiation conditions and gives information on the beam intensity monitoring. After the irradiation, the samples were etched for 17 sec in 6*M* nitric acid at room temperature to remove a 4.1–7.2 mg/cm^2 surface layer.

In the Ge(Li) γ-ray spectrum of a cast iron sample irradiated with α-particles, the 372.8 keV peak of ^{43}Sc is superimposed on the Compton continuum of a number of radionuclides produced from the matrix, and coincides with the 372.9 keV peak of ^{61}Cu. Although the interference of ^{61}Cu can be corrected for by using other peaks of

^{61}Cu, a more accurate γ-spectrometric measurement of ^{43}Sc is possible after chemical separation from the matrix activities, e.g. ^{57}Co produced by the ^{54}Fe$(\alpha,p)^{57}$Co reaction, ^{57}Ni produced by the ^{54}Fe$(\alpha,n)^{57}$Ni reaction, ^{61}Cu produced by the ^{58}Ni$(\alpha,p)^{61}$Cu reaction, and ^{66}Ga produced by the ^{63}Cu$(\alpha,n)^{66}$Ga reaction.

Table 5.24 — Irradiation conditions. (Reprinted by permission, from C. Vandecasteele, F. Alluyn, J. Dewaele and R. Dams, *Anal. Chem.*, 1985, **57**, 2549. Copyright 1985, American Chemical Society)

	Sample	Standard
α-Particle energy, MeV	17	17
Intensity, μA	3–4	0.1–0.2
Irradiation time, min	30	1–2
Monitor foil	Cu 6.9 mg/cm^2	Cu 6.9 mg/cm^2
Nuclear reaction	^{63}Cu$(\alpha,n)^{66}$Ga	
Induced activity, $t_{1/2}$	9.5 hr	
E_γ, keV	1039.4	
Additional Al foils	1.56 mg/cm^2	4.15 mg/cm^2 (series 1)
		5.72 mg/cm^2 (series 2)
		7.28 mg/cm^2 (series 3)
Effective incident energy, MeV	13.5–14.2	13.5–14.4
Residual range, mg/cm^2	34–37	

Complete dissolution of the sample, including graphite and carbides, requires approximately 5 hr of refluxing in a mixture of concentrated nitric and perchloric acids. The iron matrix can, however, also readily be dissolved in 6*M* nitric acid with heating, but a residue of graphite and carbides remains undissolved. This procedure was preferred since it is much faster. Since part of the ^{43}Sc is in the residue, this is filtered off and measured separately. The matrix activity of the residue is relatively low, allowing an easy determination of the 372.8 keV peak, which must, however, be corrected for the contribution from ^{61}Cu.

To separate scandium from radioisotopes of cobalt, nickel, copper, and gallium, cation-exchange on Dowex 50W × 8 in 1*M* hydrofluoric acid–0.1*M* nitric acid was applied. By use of ^{46}Sc $(t_{1/2} = 83.6\,d)$ tracer it was found that about 97% of the scandium is eluted from a Dowex 50W ×8 column (2.3 cm internal diameter, 12 cm height of resin bed) with 100 ml of this eluent. Since scandium is not quantitatively recovered, ^{46}Sc tracer is added before the separation, to allow determination of the chemical yield and a subsequent correction. In a Ge(Li) γ-ray spectrum of the eluate for an irradiated cast iron sample, the Compton continuum at the 372.8 keV peak is 2 orders of magnitude lower than in the spectrum obtained without separation, and no interference from ^{61}Cu occurs.

The experimental procedure was as follows. Dissolve the sample by heating in 25 ml of 6*M* nitric acid. Add 5 ml of a ^{46}Sc tracer solution and 2.5 mg of scandium(III) oxide carrier. Filter off the residue on a filter crucible (G2) and wash with 6*M* nitric acid. Evaporate to dryness and dissolve the residue in 10 ml of 1*M* hydrofluoric acid–0.1*M* nitric acid with heating. Put the solution on top of a column (2.3 cm internal diameter, 30 cm height), filled to a height of 12 cm with Dowex 50W ×8 cation-exchanger. Elute with 1*M* hydrofluoric acid–0.1*M* nitric acid and collect 100 ml of eluate. The procedure takes 4–4.5 hr.

Five ml of [46]Sc tracer solution were diluted to 100 ml with $1M$ hydrofluoric acid–$0.1M$ nitric acid and measured for 90 min with a Ge(Li) detector. The residue from the dissolution was measured with a Ge(Li) detector for 1 hr, 2–4 hr after the irradiation, and the eluate was measured 6–8 hr after the irradiation. The standards were measured for 10 min, 2–4 hr after the irradiation, either dissolved in 100 ml of $1M$ hydrochloric acid or placed in a filter crucible. The beam intensity monitor foils for the standards were measured for 10 min, 5–6 hr after the irradiation. By comparison of the 889.3 keV [46]Sc peak areas in the spectra of the tracer solution and of the eluate, the yield of each chemical separation was determined. In the residue both [43]Sc and [61]Cu occur. From a [61]Cu spectrum, obtained with α-irradiated nickel, the ratio of the area of the 372.9 keV peak to the sum of the areas of the 283.0 and 656.0 keV peaks was determined under the experimental conditions used. This ratio (0.0861 ± 0.0022) was used to correct for the contribution of [61]Cu to the 372.8 keV peak in the residue. Since the measuring geometries of the eluate and residue are different, two calibration curves were established. The first corresponded to the eluate and was obtained by measuring the calcium carbonate standards, dissolved in $1M$ hydrochloric acid. The second corresponded to the residue and was obtained by measuring the calcium carbonate standards, homogeneously distributed over a filter crucible.

Table 5.25 summarizes the results. For H 8–1, H 8–2 and H 11–3, the experimental standard deviation was significantly higher than expected from the counting

Table 5.25 — Calcium ($\mu g/g$) in cast iron determined by α-particle activation analysis. (Reprinted by permission, from C. Vandecasteele, F. Alluyn, J. Dewaele and R. Dams, *Anal. Chem.*, 1985, **57**, 2549. Copyright 1985, American Chemical Society)

Material	Results	Std. devn.*	Mean	Std. devn.†
LD–1	13.16	0.84	13.48	0.63
	12.74	0.72		
	13.93	0.73		
	14.07	0.77		
H7-1	0.90	0.28		
H8-1	19.33	0.78	16.1	2.8
	14.12	0.53		
	14.83	0.61		
H8-2	19.02	0.76	17.9	1.6
	16.81	0.61		
H9-1	13.31	0.57	13.28	0.04
	13.25	0.51		
H9-2	17.58	0.57	18.06	0.68
	18.54	0.72		
H9-3	37.9	1.0	38.48	0.81
	39.0	1.2		
H11-3	43.1	1.5	38.95	5.87
	34.8	1.3		
H16-1	9.98	0.82	9.98	0.01
	9.97	0.80		
H16-2	24.31	1.55	23.92	0.55
	23.53	1.13		

* Statistically expected standard deviation calculated from counting statistics for the measurement of [43]Sc (residue and eluate) and the measurement of [66]Ga in the monitor foil.
† Experimental standard deviation.

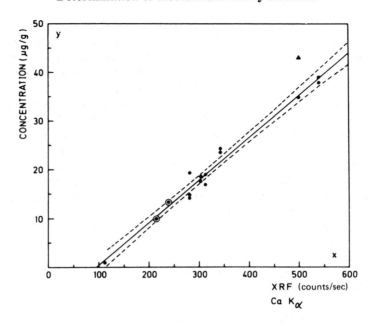

Fig. 5.3 — Calibration graph. (Reprinted by permission, from C. Vandecasteele, F. Alluyn, J. Dewaele and R. Dams, *Anal. Chem.*, 1985, **57**, 2549. Copyright 1985, American Chemical Society.)

statistics. In Fig. 5.3 the calcium concentrations for the materials from H 7–1 to H 16–2 are plotted as a function of the Ca K_α intensity: the straight line fitted to the experimental points does not pass through the origin, probably because of the presence of calcium impurities in the X-ray tube. The XRF method was calibrated by means of this straight line, which is shown in Fig. 5.3 with the 95% confidence limits for a given Ca K_α intensity.

REFERENCES

[1] J. L. Debrun, J. N. Barrandon and P. Benaben, *Anal. Chem.*, 1976, **48**, 167.
[2] J. N. Barrandon, P. Benaben and J. L. Debrun, *Anal. Chim. Acta*, 1976, **83**, 157.
[3] J. N. Barrandon, J. L. Debrun and A. Kohn, *J. Radioanal. Chem.*, 1973, **16**, 617.
[4] S. M. Kormali and E. A. Schweikert, *J. Radioanal. Chem.*, 1976, **31**, 413.
[5] K. Krivan and V. Krivan, *Z. Anal. Chem.*, 1979, **295**, 348.
[6] J. N. Barrandon, J. L. Debrun. A. Kohn and R. H. Spear, *Nucl. Instr. Meth.*, 1975, **127**, 269.
[7] S. M. Kormali, D. L. Swindle and E. A. Schweikert, *J. Radioanal. Chem.*, 1976, **31**, 437.
[8] J. L. Debrun, D. C. Riddle and E. A. Schweikert, *Anal. Chem.*, 1972, **44**, 1386.
[9] J. R. McGinley and E. A. Schweikert, *Anal. Chem.*, 1976, **48**, 429.
[10] C. Vandecasteele and J. Dewaele, Unpublished results.
[11] N. De Brucker, Unpublished results.
[12] J. L. Debrun and Ph. Albert, *Bull. Soc. Chim. France*, 1969, **3**, 1017.
[13] S. A. Dabney, D. L. Swindle, J. N. Beck, G. Francis and E. A. Schweikert, *J. Radioanal. Chem.*, 1973, **16**, 375.
[14] D. C. Riddle and E. A. Schweikert, *J. Radioanal. Chem.*, 1973, **16**, 413.
[15] J. L. Debrun and J. N. Barrandon, *J. Radioanal. Chem.*, 1973, **17**, 291.
[16] V. Krivan, D. L. Swindle and E. A. Schweikert, *Anal. Chem.*, 1974, **46**, 1626.

[17] J. N. Barrandon, P. Benaben, J. L. Debrun and M. Valladon, *Anal. Chim. Acta*, 1974, **73**, 39.
[18] P. Benaben, J. N. Barrandon and J. L. Debrun, *Anal. Chim. Acta*, 1975, **78**, 129.
[19] V. Krivan, *Anal. Chem.*, 1975, **47**, 469.
[20] J. L. Debrun, J. N. Barrandon, P. Benaben and Ch. Rouxel, *Anal. Chem.*, 1975, **47**, 637.
[21] F. Nordmann, A. Fluhr, G. Tinelli and Ch. Engelmann, *Analusis*, 1975, **3**, 171.
[22] V. Krivan, *J. Radioanal. Chem.*, 1975, **26**, 151.
[23] V. Krivan, *Anal. Chim. Acta*, 1975, **79**, 161.
[24] V. Krivan, *Talanta*, 1976, **23**, 621.
[25] S. Shibata, S. Tanaka, T. Suzuki, H. Umezawa, J. G. Lo and S. J. Yeh, *Int. J. Appl. Radiat. Isot.*, 1979, **30**, 563.
[26] P. Benaben, J. N. Barrandon and J. L. Debrun, *Radiochem. Radioanal. Lett.*, 1979, **37**, 241.
[27] W. G. Faix, J. W. Mitchell and V. Krivan, *J. Radioanal. Chem.*, 1979, **53**, 97.
[28] W. G. Faix and V. Krivan, *Talanta*, 1982, **29**, 285.
[29] C. S. Sastri and V. Krivan, *Anal. Chim. Acta*, 1982, **141**, 399.
[30] C. S. Sastri, R. Caletka and V. Krivan, *J. Trace Microprobe Tech.*, 1983, **1**, 293.
[31] R. Lacroix, G. Blondiaux, A. Giovagnoli, M. Valladon, J. L. Debrun, R. Coquille and M. Gauneau, *J. Radioanal. Nucl. Chem.*, 1984, **83**, 91.
[32] P. Goethals, B. Bosman and C. Vandecasteele, *Bull. Soc. Chim. Belg.*, 1986, **95**, 331.
[33] R. Dams, F. Alluyn, B. Vanloo, G. Wauters and C. Vandecasteele, *Z. Anal. Chem.*, 1986, **325**, 163.
[34] N. De Brucker, K. Strijckmans and C. Vandecasteele, *Anal. Chim. Acta*, 1987, **195**, 323.
[35] C. Vandecasteele, J. Dewaele, M. Esprit and P. Goethals, *Anal. Chim. Acta*, 1980, **119**, 121.
[36] W. Schmid, K. P. Egger and V. Krivan, *J. Radioanal. Nucl. Chem.*, in the press.
[37] C. S. Sastri, H. Petri and G. Erdtmann, *Anal. Chem.*, 1977, **49**, 1510.
[38] A. Giovagnoli, C. Koemmerer, M. Valladon, G. Blondiaux and J. L. Debrun, *Radiochem. Radioanal. Lett.*, 1979, **41**, 409.
[39] C. S. Sastri, H. Petri and G. Erdtmann, *J. Radioanal. Chem.*, 1977, **38**, 157.
[40] C. S. Sastri, H. Petri and G. Erdtmann, *Anal. Chim. Acta*, 1977, **89**, 141.
[41] P. Goethals, C. Vandecasteele and J. Hoste, *Anal. Chim. Acta*, 1978, **101**, 63.
[42] C. Vandecasteele, F. Alluyn, J. Dewaele and R. Dams, *Anal. Chem.*, 1985, **57**, 2549.

6

Analysis of geological, environmental and biological samples

Only a few applications of charged particle activation analysis to geological, environmental and biological samples have been described. The technique encounters some problems with powdered samples having a low thermal conductivity. For example, heating of the sample during irradiation may result in volatilization of the element of interest or of part of the matrix. Therefore, the thermal behaviour of environmental materials and rocks during proton irradiation will first be discussed in some detail.

6.1   THERMAL BEHAVIOUR OF ROCKS AND ENVIRONMENTAL SAMPLES UNDER PROTON IRRADIATION

As discussed in Section 2.3, in CPAA a relative method is usually used, in which thick samples and standards are irradiated. In the irradiated material the protons lose their energy, which is transformed into heat. The power P (watt), dissipated in the sample during irradiation with protons of energy E_I (MeV) at intensity I (μA) is given by Eq. (6.1)

$$P = E_I \frac{I}{e} \tag{6.1}$$

where e is the charge on an electron. Owing to the low thermal conductivity of the materials considered (rocks, fly-ash, sediments, etc.) high temperatures are reached. In addition, these samples are usually powders. On irradiation of a fly-ash sample under vacuum and under normal irradiation conditions, temperatures in excess of 1000°C may be reached, which may lead to volatilization of some components. When the analyte element or the radionuclide produced is volatile, low results are obtained. When the analyte element and the indicator nuclide are not volatile, but some matrix components are, a positive error arises since the analyte element is enriched in the irradiated part of the sample.

The latter effect was noted by Xenoulis *et al.* [1] during the determination of light elements in biological samples by proton-induced prompt-gamma emission analysis (PIGE). This method allows study of the stability of a material under irradiation with charged particles: the yield of prompt γ-rays from a given nuclear reaction is measured as a function of time. For the $^{23}Na(p, p'\gamma)^{23}Na$ and $^{24}Mg(p, p'\gamma)^{24}Mg$ reactions, the yields of prompt γ-rays increase by 33 and 30% respectively for lyophilized plant materials irradiated with a 10 nA beam of 4.3 MeV protons. This increase can be explained by volatilization of part of the matrix, due to heating of the sample during irradiation. As soon as an equilibrium temperature is reached, the yield remains constant. In another experiment, where the plant material was ashed at 300°C before the irradiation and irradiated under identical conditions, no increase in the yield was observed.

Environmental samples are usually mainly composed of non-volatile silicates and may contain a non-negligible fraction of organic or volatile components. The thermal behaviour of the sample during the irradiation must thus be considered. In order to obtain low detection limits, relatively high beam intensities, up to $1 \mu A$, must be used. In most irradiation facilities for activation analysis with charged particles, the samples are irradiated under vacuum. Because of their low thermal conductivity, the materials considered must be irradiated in a medium with a higher thermal conductivity than a vacuum. Zikovsky and Schweikert [2] mixed biological samples with pure graphite and pressed it into a pellet, in order to increase the thermal conductivity. Disadvantages of this method are the higher detection limits and increased risks of contamination.

Wauters *et al.* [3] irradiated the samples under helium instead of under vacuum, using the experimental set-up described in Section 3.1.2. In this system the temperature during the irradiation was measured by means of a thermocouple, for environmental samples and rocks irradiated with 23 MeV protons. Figure 3.8 (p. 61) shows the experimental arrangement. A small NiCr–Ni thermocouple was placed along the axis of the disc-shaped sample and connected to a sintered polyethylene disc through which the helium gas diffused. The thermocouple was connected to a recorder, to register the potential differences, from which the temperature was deduced. After a short irradiation time an equilibrium temperature was reached.

The certified reference material NBS-SRM 1633a, Coal Fly Ash, was irradiated with 23 MeV protons under different irradiation conditions and the temperature was recorded until thermal equilibrium was reached.

6.1.1 Temperature as a function of depth in the sample

The amount of energy lost by a proton beam per unit distance of penetration depends on the depth in the sample, since the stopping power increases with decreasing proton energy. This is shown in Fig. 6.1. If it is assumed that no heat transport along the axis of the sample occurs, the temperature reached during the irradiation will show the same course as the energy loss: the highest temperature is reached around the end of the range.

Figure 6.2 gives the measured temperature for an irradiation under vacuum at 200 nA and under helium (3 bar) at intensities of 200, 500 and 1000 nA, respectively. On the abscissa 100 mg/cm² corresponds to a depth of ca. 1 mm in the fly-ash. By varying the amount of fly-ash in the sample holder, it was possible to carry out

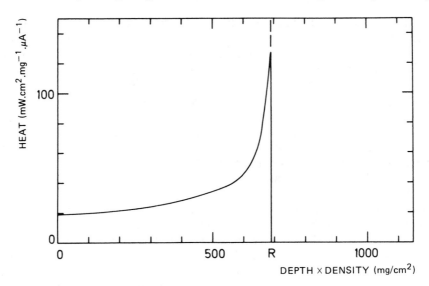

Fig. 6.1 — Energy lost by 23 MeV protons in NSB-SRM 1633a, Coal Fly Ash. (Reproduced by permission, from G. Wauters, C. Vandecasteele and J. Hoste, *J. Radioanal. Nucl. Chem. Articles,* 1986, **98**, 345. Copyright, 1986, Akadémiai Kiadó, Budapest.)

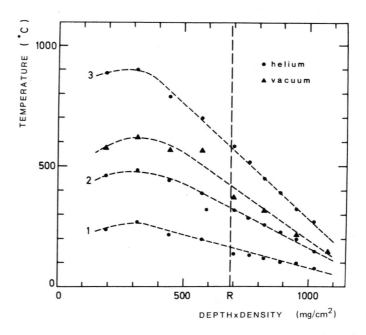

Fig. 6.2 — Temperature curves as a function of depth, for NBS-SRM 1633a, Coal Fly Ash, irradiated with 23 MeV protons: ▲ under vacuum at 200 nA, ● under helium (3 bar), at 200 nA (curve 1), 500 nA (curve 2) and 1000 nA (curve 3) intensity. (Reproduced by permission, from G. Wauters, C. Vandecasteele and J. Hoste, *J. Radioanal. Nucl. Chem. Articles,* 1986, **98**, 345. Copyright, 1986; Akadémiai Kiadó, Budapest.)

measurements at different positions, both before and after the proton range. The temperature distributions all have the same shape, with a maximum at around half the range, indicating that heat is transferred along the axis of the sample. For the same intensity, temperatures reached under vacuum are higher than those under helium, and, as expected, the temperature increases with the beam intensity. Under vacuum, even at low intensities (200 nA), a temperature in excess of 600°C is reached; under helium (3 bar) at the same beam intensity, the temperature is around 260°C.

6.1.2 Temperature as a function of the helium pressure
Figure 6.3 gives, for an irradiation at 500 nA intensity, the maximum temperature as

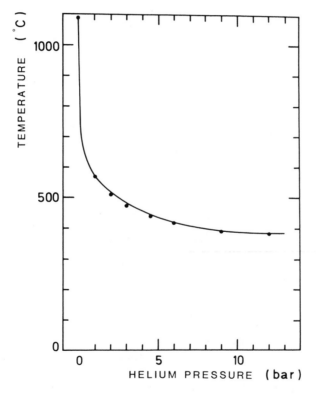

Fig. 6.3 — Temperature as a function of the helium pressure in the sample holder, for NBS-SRM 1633a, irradiated with 23 MeV protons at 500 nA intensity. (Reproduced by permission, from G. Wauters, C. Vandecasteele and J. Hoste, *J. Radioanal. Nucl. Chem. Articles,* 1986, **98**, 345. Copyright, 1986, Akadémiai Kiadó, Budapest.)

a function of the helium pressure in the sample holder. The curve shows a rapid decrease in the beginning, when going from vacuum to 1 bar, followed by a slow decrease at higher pressures. Above 9 bar the temperature does not change significantly, so this pressure is used in practice.

6.1.3   Temperature as a function of the beam intensity

The maximum temperature for an irradiation under 9 bar helium pressure at different intensities was determined. Figure 6.4 shows the results. At 200 nA, for

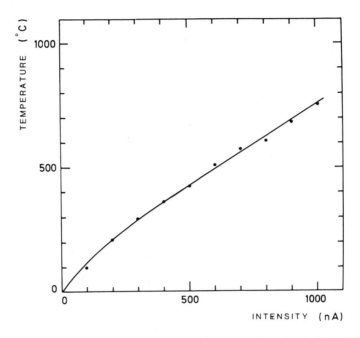

Fig. 6.4 — Maximum temperature as a function of the beam intensity for NBS-SRM 1633a, irradiated with 23 MeV under helium (9 bar). (Reproduced by permission, from G. Wauters, C. Vandecasteele and J. Hoste, *J. Radioanal. Nucl. Chem. Articles*, 1986, **98**, 345. Copyright, 1986, Akadémiai Kiadó, Budapest.)

instance, a temperature of 210°C is reached, whereas for an irradiation under vacuum 610°C is attained.

Whether this curve is also valid for other environmental samples depends on the thermal conductivity of the samples and on the grain size of the powder. As a check, the maximum temperature was determined for a number of reference materials with different compositions. Under 9 bar pressure of helium, at an irradiation intensity of 500 nA the maximum temperature was 430°C for NBS SRM 688, basalt Rock, 380°C for BCR CRM 142, Light Sandy Soil, and 380°C for BCR CRM 176, City Waste Incineration Ash. These values are in good agreement with the maximum temperature of 420°C for NBS SRM 1633a. The curve given in Fig. 6.4 can thus generally be used to estimate the maximum temperature in an environmental sample irradiated with 23 MeV protons under helium at a pressure of 9 bar.

6.1.4   Thermogravimetry

Figure 6.5 gives the results of a thermogravimetric analysis of the samples considered in Section 6.1.3. It is clear that below 1000°C no loss of mass occurs for the basalt rock (SRM 688), whereas for the soil sample (CRM 142) and the incineration ash (CRM 176) losses occur already at 100°C. The fly-ash (SRM 1633a) behaves intermediately.

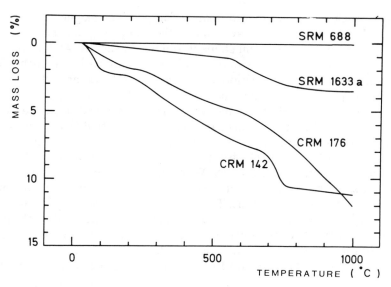

Fig. 6.5 — Thermogravimetric curves. (Reproduced by permission, from G. Wauters, C. Vandecasteele and J. Hoste, *J. Radioanal. Nucl. Chem. Articles,* 1986, **98**, 345. Copyright, 1986, Akadémiai Kiadó, Budapest.)

For a given beam intensity the loss of mass during the irradiation can be deduced from Figs. 6.4 and 6.5. The loss of mass thus estimated is, however, a maximum value, since the temperature given in Fig. 6.4 is reached in only a small part of the sample, and the curves in Fig. 6.5 correspond to the thermal behaviour in an open system under 1 bar pressure, whereas the samples are irradiated in a closed system under helium at 9 bar pressure.

For analysis where a low beam intensity yields a sufficient sensitivity, the beam intensity can be chosen so that no loss of mass occurs during the irradiation. Irradiation of SRM 1633a at 500 nA yields a maximum temperature of 400°C, so the mass loss is less than 1%. Irradiation of CRM 142 under the same conditions would yield a mass loss of at most 5%, so a lower intensity is preferable. At 100 nA the loss of mass is around 2%. When, for samples with low thermal stability, high intensities are necessary, the sample can be heated before the irradiation. The final analytical result must then of course be corrected for the loss of mass during heating.

For irradiations with ^{3}He or ^{4}He much higher temperatures are reached in the sample for the same beam intensity. These particles are therefore rarely used for the analysis of geological, environment and biological samples.

6.2 STANDARDIZATION

To calculate the analytical results, either by numerical integration or by approximate methods as described in Section 2.3, the composition of the sample and the standard must, in principle, be known.

Matrices such as rock or fly-ash, mainly composed of magnesium, aluminium, calcium and iron silicates, all have almost the same stopping power, since the stopping powers of the component elements differ only slightly [4]. For this reason it can be expected that $F(E_I)$ (see Section 2.3) for such matrices is not much influenced by variations of the bulk composition of the sample. Problems arise, however, when organic components are present in the sample, as is, for instance, the case for some sediments, sludges and soils. Since the stopping powers of hydrogen and carbon are respectively ca. 3 and 1.2 times that of any of the elements that form a silicate matrix, the stopping power of these matrices will be significantly higher than that of a silicate matrix. Exact knowledge of the hydrogen and carbon concentrations is thus required to calculate an accurate $F(E_I)$ value. When the concentration of hydrogen and carbon in the sample is not sufficiently well known, it may be necessary to remove the organic fraction before the analysis. This can be done by low-temperature ashing or by heating the material to 450°C in air. $F(E_I)$ can then be calculated for the inorganic residue, which is mainly composed of silicates. The analytical result must, of course, be corrected for the loss of mass, in order to relate it to the original sample.

Table 6.1 [4] gives $F(E_I)$ values for the determination of lead and cadmium in

Table 6.1 — $F(E_I)$ values for some geological and environmental reference materials. (Reproduced from G. Wauters, C. Vandecasteele, K. Casteleyn and K. Strijckmans, *J. Trace Microprobe Tech.*, 1987, **5**, 169, by courtesy of Marcel Dekker, Inc.)

Reference material	Loss of mass	$F(E_I)$ 22.3 MeV/PbO	$F(E_I)$ 14.1 MeV/CdO
NBS-SRM 688 Basalt Rock	—	1.774	1.461
USGS PCC-1 Peridotite Rock	—	1.809	1.493
USGS DTS-1 Dunite Rock	—	1.789	1.474
BCR-CRM 176 City Waste Incineration Ash	—	1.746	1.437
NBS-SRM 1633a Coal Fly Ash	—	1.767	1.456
NBS-SRM 1645 River Sediment	5.3%	1.754	1.445
BCR-CRM 146 Sewage Sludge	26.2%* 29.6%†	1.772	1.460

* Low-temperature ashing.
† Heating in air at 450°C.

several certified reference materials with well known matrix compositions, calculated by numerical integration. For lead, 22.3 MeV protons and a lead(II) oxide standard are used, for cadmium 14.5 MeV protons and a cadmium(II) oxide standard. For NBS-SRM 1645 and BCR-CRM 146 the loss of mass during ashing is indicated and $F(E_I)$ was calculated for the inorganic residue. The $F(E_I)$ values in Table 6.1 differ by less than 4%.

For other materials not mainly composed of silicates, for instance coal, the matrix composition must be known in order to allow the calculation of $F(E_1)$. When the matrix composition is not sufficiently well known, an obvious solution consists in the use of an internal standard. For this the sample concentration of an element (other than the analyte) that is easily determined by charged particle activation must be known, for instance, from very accurate and precise analysis by another analytical technique; alternatively, a known amount of the internal standard may be mixed with the sample. When the sample of interest and a standard with known composition containing the element used as internal standard, are irradiated with particles of the same energy and the activities (a_X, a_S) for the internal-standard element are measured, the range R_X in the sample can be deduced from

$$c_X = c_S \frac{a_X}{i_X(1 - e^{-\lambda t_{b,X}})} \frac{i_S(1 - e^{-\lambda t_{b,S}})}{a_S} \frac{R_S}{R_X} \qquad (6.2)$$

which is a combination of Eqs. (2.49) and (2.55). Equation (6.2) is valid with the same restrictions as Eq. (2.55) (see Section 2.3.4). The concentration of the analyte element in the sample can then be calculated from the activities a_X and a_S for the analyte in the sample and standard, by means of Eq. (6.2). Advantages of the internal standard method are that

(1) the composition of the sample need not be known;
(2) partial volatilization of the matrix does not lead to systematic errors, provided that neither the analyte element not the internal-standard element is volatile.

6.3 APPLICATIONS

Analyses of environmental and geological materials fall into two groups.

(1) In most case the highest possible accuracy is not required, but a large number of samples must be analysed.
(2) Sometimes, for instance for certification analysis of reference materials, the highest possible accuracy is required. In this case all possible sources of systematic error must be studied and taken into account. In order to obtain low detection limits and to measure the induced activity with the best possible precision, it may sometimes be necessary to develop radiochemical separations for the radionuclides of interest.

Table 6.2 gives a literature survey of the analysis of geological, environmental and biological samples by proton activation analysis.

Examples of methods from groups 1 (Section 6.3.1) and 2 (Sections 6.3.2 and 6.3.3) will be discussed in more detail.

Table 6.2 — Literature survey of the analysis of geological, environmental and biological samples by proton activation analysis.

Reference	Matrix	Element determined
Landsberger et al. [5]	Coal	S
Bonardi et al. [6]	Biological	Heavy metals
Delmas et al. [7]	Minerals	13 elements
Desaedeleer and Ronneau [8]	Airborne particulate matter	Pb
Desaedeleer et al. [9]	Airborne particulate matter	Pb
Strijckmans et al. [10]	Fly-ash	S
Wauters et al. [11]	Environmental Rocks	Pb
Wauters et al. [12]	Environmental	Cd, Tl, Pb
Wauters et al. [4]	Environmental	13 elements
Zikovsky and Schweikert [2]	Botanical	10 elements
	Animal tissue	7 elements
Priest et al. [13]	Airborne particulate matter	18 elements
Parsa and Markowitz [14]	Airborne particulate matter	Pb

6.3.1   Determination of lead in airborne particulate matter by instrumental proton activation analysis

Desaedeleer et al. [9] describe a method for the determination of lead, in airborne particulate matter collected on filters, by irradiation with 50 MeV protons. $^{204}Bi(t_{1/2}$ = 11.22 hr) is produced from lead by the $^{206}Pb(p, 3n)^{204}Bi$, $^{207}Pb(p, 4n)^{204}Bi$ and $^{208}Pb(p, 5n)^{204}Bi$ reactions. Because of its shorter half-life, ^{204}Bi is more suitable than $^{206}Bi(t_{1/2} = 6.243$ d) for the rapid determination of minute amounts of lead. The half-life is, however, long enough for a large number of samples — up to 40 — to be irradiated simultaneously, and subsequently counted, making the method suitable for routine application.

As standards, thin aluminium foils (2.7 mg/cm^2) covered with lead (2 mg/cm^2) by deposition in vacuum, or Whatman 41 cellulose filter paper impregnated with a lead nitrate solution, were used. Typical samples consisted of 1–10 μg/cm^2 of aerosol collected on Whatman or Sartorius filter paper. In the procedure targets were covered on both sides with thin foils of high-purity aluminium and sandwiched between aluminium rings with a hole drilled through them to allow circulation of helium to cool the targets. Irradiations were carried out with 50 MeV protons at 1 μA intensity for 30 min. Several standards and samples were irradiated together.

The samples were measured with a germanium semiconductor detector. The 374.7 keV γ-ray of ^{204}Bi was mostly used, but the 899.3 keV γ-ray was sometimes more appropriate. Typical counting times were 5 min for standards and 5–50 min for samples. Quantitative analysis was made by comparison of the activities in the sample and in the set of standards. Energy loss in the stack of samples and standards could, to a first approximation, be neglected.

The accuracy of the method was tested by analysing the same samples by proton activation analysis and by ETA–AAS. The agreement was usually fairly good, except for low lead concentrations. The precision (same filter analysed at 6 different points) was better than ± 15%.

The method was applied for routine determination of lead in airborne particulate matter. The purpose was to determine the contribution of freeway automobile emissions to ambient lead concentrations in a rural area. The method is particularly sensitive and allows the determination of ng amounts of lead on a $1\,cm^2$ filter area. Lead can thus be determined in aerosol samples from less than $0.1\,m^3$ of air from remote areas.

In a later paper Priest et al. [13] showed that activation with 30 MeV protons allows the instrumental multi-element analysis of airborne particles collected on filter paper. Fifteen–twenty samples were stacked and irradiated for 2 hr with a 0.3 μA beam of 30 MeV protons. The samples were measured with a Ge(Li) spectrometer 4 hr, 15 hr and 10 days after the irradiation. (Sodium), magnesium, (chlorine), (calcium), (titanium), (chromium), (manganese), iron, (nickel), (copper), zinc, (arsenic), (bromine), (strontium), (cadmium), (tin), (antimony) and lead could be determined in typical urban aerosols with a sampled volume of 1 (10) m^3 of air (as indicated by the parentheses).

6.3.2   Instrumental proton activation analysis of solid environmental samples

To evaluate the possibilities of instrumental proton activation analysis for the precise and accurate analysis of environmental samples available as powdered samples, Wauters et al. [4] analysed BCR-CRM 176, City Waste Incineration Ash, a reference material with well known matrix composition and certified for several minor and trace elements. For this, 12 MeV protons were used, to obtain a good compromise between sensitivity and freedom from nuclear interferences [due, for instance, to (p, d) and (p, α) reactions].

Table 6.3 summarizes the most important nuclear reactions induced in environmental materials by 12 MeV protons, along with the threshold energy and nuclear data for the radionuclides produced. The standards used are given in Table 6.4.

The samples and the standards were irradiated under helium (9×10^5 Pa) in a water-cooled sample holder. The beam intensity was 100 nA for the samples and the irradiation time was 10 min for short-lived radionuclides and 1 hr for long-lived radionuclides. Before the sample and the standards a 12.5 μm nickel foil was placed to serve as a beam intensity monitor. The ^{55}Co activity ($t_{1/2} = 17.54$ hr; $E_\gamma = 931.5$ keV) from the ^{58}Ni$(p, \alpha)^{55}$Co reaction was used as a measure of the beam intensity.

The samples, the standards and the nickel foils were measured with a germanium semiconductor detector. Short-lived radionuclides were measured for 1 hr after a decay time of 1–6 hr and long-lived radionuclides for 4 hr after a 1–3 d decay time.

Table 6.5 summarizes possible nuclear interferences. The interferences were determined experimentally and are also given in Table 6.5. in general, the interferences were negligible owing to the small interference factor or to the low concentration of the interfering element. Only for the determination of calcium by the reactions leading to ^{44}Sc and ^{44m}Sc was the interference from titanium significant, and a correction applied, amounting to 0.58 and 7.0%, respectively.

Table 6.6 gives the results and the detection limits. For calcium, chromium, zinc and tin, the results obtained for different experimental conditions and by using different nuclear reactions, are in excellent agreement. Except for calcium and tin, the agreement between the results of Wauters et al. [4] and the certified or indicative

Table 6.3 — Nuclear reactions induced in environmental materials by 12 MeV protons. (Reproduced from G. Wauters, C. Vandecasteele, K. Casteleyn and K. Strijckmans, *J. Trace Microprobe Tech.*, 1987, **5**, 169, by courtesy of Marcel Dekker, Inc.)

Element	Nuclear reaction	Threshold energy, MeV	Half-life	γ-Ray energies, keV (absolute intensity, %)
Ca	^{44}Ca(p,n)^{44}Sc	4.5	3.93 hr	1157.0 (100)
	^{44}Ca(p,n)^{44m}Sc	4.5	2.44 d	271.4 (86)
Ti	^{48}Ti(p,n)^{48}V	4.9	16.1 d	944.3 (8); 983.5 (100); 1311.6 (98)
Cr	^{52}Cr(p,n)^{52m}Mn	5.6	21.3 min	1434.3 (100)
	^{52}Cr(p,n)^{52}Mn	5.6	5.7 d	744.2 (85)
Fe	^{56}Fe(p,n)^{56}Co	5.5	77.3 d	1238.3 (68); 1771.5 (16)
Cu	^{63}Cu(p,n)^{63}Zn	4.2	38.1 min	962.1 (7)
	^{65}Cu(p,n)^{65}Zn	2.1	243.8 d	1115.5 (51)
Zn	^{64}Zn(p,α)^{61}Cu	0.8	3.41 hr	283.0 (13); 656.0 (10); 1185.7 (4)
	^{68}Zn(p,n)^{68}Ga	3.8	67.8 min	1077.4 (3)
	^{67}Zn(p,n)^{67}Ga	1.8	78.2 hr	300.2 (15); 393.6 (4)
Sr	^{87}Sr(p,n)^{87}Y	2.7	80.3 hr	388
Y	^{89}Y(p,n)^{89}Zr	3.6	78.4 hr	909.1 (100)
Zr	^{90}Zr(p,n)^{90}Nb	7.0	14.6 hr	141.2 (67); 1129.1 (92)
Cd	^{111}Cd(p,n)^{111}In	1.6	2.83 d	245.4 (94)
Sn	^{118}Sn(p,n)^{118m}Sb	4.5	5.0 hr	253.7 (96); 1229.6 (100)
	^{122}Sn(p,n)^{122}Sb	2.4	2.70 d	564.1 (71)
Pb	^{206}Pb(p,n)^{206}Bi	4.4	6.243 d	881.0 (68); 1718.7 (34)

Table 6.4 — Standards. (Reproduced from G. Wauters, C. Vandecasteele, K. Casteleyn and K. Strijckmans, *J. Trace Microprobe Tech.*, 1987, **5**, 169, by courtesy of Marcel Dekker, Inc.)

Element	Standard	Element	Standard
Ca	$CaCO_3$	Sr	$Sr(NO_3)_2$
Ti	TiO_2	Y	Y_2O_3
Cr	K_2CrO_4	Zr	ZrN
Fe	Fe	Cd	CdO
Cu	CuO	Sn	SnO_2
Zn	ZnO	Pb	$PbSO_4$

Table 6.5 — Nuclear interferences and interference factors for 12 MeV protons. (Reproduced from G. Wauters, C. Vandecasteele, K. Casteleyn and K. Strijckmans, *J. Trace Microprobe Tech.*, 1987, **5**, 169, by courtesy of Marcel Dekker, Inc.)

Element	Nuclear reaction (Threshold energy, MeV)		Interference reaction (Threshold energy, MeV)		Interference factor
Ca	$^{44}Ca(p,n)^{44}Sc$	(4.5)	$^{45}Sc(p,d)^{44}Sc$	(9.3)	$< 8.23 \times 10^{-4}$
			$^{47}Ti(p,\alpha)^{44}Sc$	(2.3)	5.16×10^{-2}
	$^{44}Ca(p,n)^{44m}Sc$	(4.5)	$^{45}Sc(p,d)^{44m}Sc$	(9.3)	$< 5.65 \times 10^{-3}$
			$^{47}Ti(p,\alpha)^{44m}Sc$	(2.3)	6.36×10^{-1}
Cu	$^{63}Cu(p,n)^{63}Zn$	(4.2)	$^{64}Zn(p,d)^{63}Zn$	(9.9)	$< 2.10 \times 10^{-2}$
	$^{65}Cu(p,n)^{65}Zn$	(2.1)	$^{66}Zn(p,d)^{65}Zn$	(9.0)	$< 9.99 \times 10^{-4}$
			$^{69}Ga(p,\alpha n)^{65}Zn$	(8.7)	$< 6.02 \times 10^{-4}$
Zn	$^{68}Zn(p,n)^{68}Ga$	(3.8)	$^{69}Ga(p,d)^{68}Ga$	(8.2)	$< 1.02 \times 10^{-2}$
			$^{72}Ge(p,\alpha n)^{68}Ga$	(8.8)	$< 1.16 \times 10^{-2}$
	$^{64}Zn(p,\alpha)^{61}Cu$	(0.8)	$^{61}Ni(p,n)^{61}Cu$	(3.0)	2.49×10^{-1}
	$^{67}Zn(p,n)^{67}Ga$	(1.8)	$^{70}Ge(p,\alpha)^{67}Ga$	$(Q>0)$	1.79×10^{-1}
Sr	$^{87}Sr(p,n)^{87}Y$	(2.7)	$^{90}Zr(p,\alpha)^{87}Y$	(0.9)	$< 6.29 \times 10^{-2}$
			$^{91}Zr(p,\alpha n)^{87}Y$	(8.2)	
Y	$^{89}Y(p,n)^{89}Zr$	(3.6)	$^{90}Zr(p,d)^{89}Zr$	(9.9)	$< 1.00 \times 10^{-3}$
			$^{93}Nb(p,\alpha n)^{89}Zr$	(5.7)	$< 6.72 \times 10^{-5}$
Zr	$^{90}Zr(p,n)^{90}Nb$	(7.0)	$^{94}Mo(p,\alpha n)^{90}Nb$	(9.1)	$< 9.95 \times 10^{-5}$
Cd	$^{111}Cd(p,n)^{111}In$	(1.6)	$^{114}Sn(p,\alpha)^{111}In$	$(Q>0)$	$< 2.60 \times 10^{-4}$
			$^{112}Sn(p,2p)^{111}In$	(7.6)	
Sn	$^{122}Sn(p,n)^{122}Sb(2.4)$		$^{123}Sb(p,d)^{122}Sb$	(6.9)	1.00×10^{-2}
			$^{123}Te(p,2p)^{122}Sb$	(8.2)	$< 4.61 \times 10^{-3}$

values given by BCR is also satisfactory. For calcium and tin, however, several values which differ significantly are given in the certification report.

When the samples were irradiated under vacuum at 500 nA, instead of under helium at 100 nA, the results obtained were: 933 ± 33 $\mu g/g$ for chromium, 24200 ± 400 $\mu g/g$ for iron, 28320 ± 840 $\mu g/g$ for zinc and 619 ± 41 $\mu g/g$ for cadmium, all significantly higher than those given in Table 6.6. This illustrates clearly the effectiveness of irradiating under helium with the sample holder described in Section 3.1.2.

Instrumental proton activation analysis allows the determination of some 12

Table 6.6 — Results (μg/g) for BCR-CRM 176 obtained by instrumental proton activation analysis. (Reproduced from G. Wauters, C. Vandecasteele, K. Casteleyn and K. Strijckmans, *J. Trace Microprobe Tech.*, 1987, **5**, 169, by courtesy of Marcel Dekker, Inc.)

Ele-ment	Ref. [4]				Certified (C) or indicative (I) value; 95% confidence interval		Detection limit
	$\bar{x}$	s	n				
Ca	89200	4100	6	b	81000 ± 5000; 87900 ± 500	I	660
	92000	3400	3	d			2500
Ti	10100	280	3	d	9800 ± 700; 10000 ± 1000	I	10
Cr	850	20	6	a	863 ± 30	C	35
	850	32	3	d			43
Fe	21970	210	3	d	21300 ± 1100	C	120
Cu	1270	70	6	a	1302 ± 26	C	470
Zn	26400	1600	6	b	25770 ± 380	C	3400
	25000	1400	6	a			1300
	26410	390	3	d			320
Sr	457	17	3	d	433 ± 20	I	51
Y	18	2	3	d			5
Zr	149	9	3	c			7
Cd	477	11	3	d	470 ± 9	C	47
Sn	8150	980	6	b	6670 ± 50; 4700 ± 100;		3000
	7980	90	3	d	5400 ± 100	I	270
Pb	11610	440	3	d	10870 ± 170	C	250

x: mean value of n analysis. s: standard deviation. a: by short-lived radionuclide after 1 hr decay time. b: by short-lived radionuclide after 6 hr decay time. c: by long-lived radionuclide after 1 d decay time. d: by long-lived radionuclide after 3 d decay time.

elements in fly-ash. Some of these are difficult (calcium, titanium, copper, strontium and tin) or even impossible (yttrium, zirconium and lead) to determine by instrumental neutron activation analysis.

6.3.3 Determination of cadmium, thallium and lead in environmental samples by radiochemical proton activation analysis [12]

It appears from Table 6.6 that detection limits for elements such as cadmium and lead, by instrumental proton activation analysis, are rather high. Cadmium, thallium and lead are, however, all of major environmental interest. When lower detection limits are required, higher proton energies (15–23 MeV) are used and a post-irradiation radiochemical separation is required to separate the radionuclides of interest from the matrix activities.

Table 6.7 gives the most important proton-induced nuclear reactions of cadmium, thallium and lead, along with nuclear data on the radionuclides produced and nuclear interference reactions.

For the determination of cadmium by the ^{111,112,113}Cd(p, xn)^{111}In nuclear reactions, nuclear interference from tin and indium must be considered. These interferences can be avoided by limiting the incident energy to below the threshold energy of the interfering reactions.

For the determination of thallium the ^{203,205}Tl(p, xn)^{203}Pb and ^{203}Tl(p, 3n)^{201}Pb reactions can be used. For the former reactions nuclear interference from lead must be taken into account, the correction being important when the lead concentration is much higher than the thallium concentration. In addition, the correction is somewhat complicated, since ^{203}Pb is formed directly from ^{204}Pb and also through the

Table 6.7 — Nuclear data for the determination of cadmium, thallium and lead with 23 MeV protons. (Reproduced by permission, from G. Wauters, C. Vandecasteele, K. Strijckmans and J. Hoste, *J. Radioanal. Nucl. Chem. Articles*, 1987, **112**, 23. Copyright 1987, Akadémiai Kiadó, Budapest.)

Nuclear reaction (Threshold energy, MeV)	Half-life	γ-Ray energy, keV	Nucelar interference reaction (Threshold energy, MeV)
^{111}Cd(p, n)^{111}In(1.6)	2.83 d	171.3	^{116}Sn(p, α2n)^{111}In(14.5)
^{112}Cd(p, 2n)^{111}In(11.1)		245.4	^{117}Sn(p, α3n)^{111}In(21.4)
^{113}Cd(p, 3n)^{111}In(17.7)			^{113}In(p, p2n)^{111}In(17.2)
^{203}Tl(p, n)^{203}Pb(1.8)	2.17 d	279.2	^{204}Pb(p, pn)^{203}(8.5)
^{205}Tl(p, 3n)^{203}Pb(16.1)			^{204}Pb(p, 2n)^{203}Bi $\xrightarrow{\beta+}$ ^{203}Pb(12.5)
^{203}Tl(p, 3n)^{201}Pb(17.8)	9.4 hr	331.2	
^{206}Pb(p, n)^{206}Bi(4.4)	6.243 d	803.1	^{209}Bi(p, p3n)^{206}Bi(22.4)
^{207}Pb(p, 2n)^{206}Bi(11.3)		881.0	
^{208}Pb(p, 3n)^{206}Bi(18.6)			

decay of ^{203}Bi. Therefore the ^{203}Tl(p, 3n)^{201}Pb reaction, which is interference-free, is preferred.

For a proton energy below 22.4 MeV, the determination of lead by using the $^{206, 207, 208}$Pb(p, xn)^{206}Bi reactions is free from nuclear interferences.

Two certified reference materials were analysed: NBS-SRM 1633a, Coal Fly Ash, and BCR-CRM 176, City Waste Incineration Ash, and 0.8 g of sample was used per analysis. As standards, thick targets of cadmium(II) oxide, thallium(III) oxide and lead(II) oxide were used.

The samples and the standards were irradiated with 23 or 15 MeV protons under helium in a watercooled sample holder (Fig. 3.7, p. 60). A 50 μm titanium foil separated the helium from the vacuum in the beam transport system, and before the sample and the standard a 12.5 μm nickel foil was placed, serving as a beam intensity monitor. The activity of ^{57}Ni($t_{1/2}$ = 36.0 hr; E_γ = 1377.6 keV) from ^{58}Ni(p, pn)^{57}Ni was used as a measure of the beam intensity. The foils degraded the proton energy from 23.0 to 22.4 and from 15.0 to 14.1 MeV. BCR-CRM 176 was irradiated for 1 hr at a beam intensity of 250 nA, NBS-SRM 1633a for 1 hr at 500 nA, the standards for a few minutes at 50 nA.

High concentrations of cadmium ($>$ 100 μg/g) and lead ($>$ 500 μg/g) could be determined instrumentally. For lower concentrations a post-irradiation chemical separation of ^{111}In, ^{201}Pb and ^{206}Bi from radionuclides produced from the matrix was required. The samples were fused with a sodium carbonate–sodium hydroxide–sodium nitrate mixture and the cooled melts were dissolved in hydrochloric acid. The solution was brought onto a column of Dowex 1×8 anion-exchanger in the chloride form. The interfering radionuclides were eluted with 2M hydrochloric acid, indium, lead and bismuth being retained on the column. Indium and lead were eluted with 6M nitric acid, and the bismuth was eluted with 0.5M sulphuric acid. Figure 6.6 shows the elution curves for ^{56}Co, ^{48}V, ^{52}Mn, ^{44m}Sc, ^{67}Ga, ^{111}In, ^{203}Pb, ^{62}Zn and ^{206}Bi, obtained by treating as decribed a solution of an inactive fly-ash sample (NBS-SRM 1633a), to which these radionuclides were added.

The procedure for the chemical separation is as follows. To the irradiated sample in a nickel crucible add 4 g of sodium carbonate, 2.5 g of sodium hydroxide

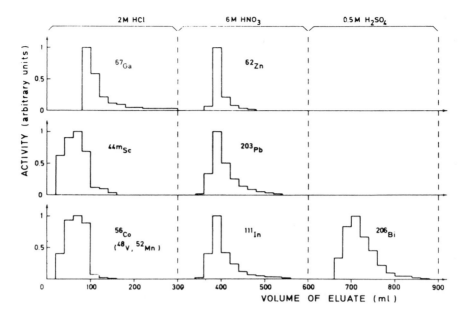

Fig. 6.6 — Elution curves. (Reproduced by permission, from G. Wauters, C. Vandecasteele,
K. Strijckmans and J. Hoste, *J. Radioanal. Nucl. Chem. Articles,* 1987, **112**, 23. Copyright
1987, Akadémiai Kiadó, Budapest.)

and 1.5 g of sodium nitrate, mix, and heat with a Bunsen burner until fusion is
complete. Dissolve the cooled melt in 100 ml of 2*M* hydrochloric acid and add 1 ml of
a solution of 400 μg/ml each of indium(III) oxide, lead(II) oxide and bismuth(III)
nitrate in 2*M* hydrochloric acid (to provide a carrier). Filter off the silica on a glass
filter and evaporate the filtrate to 60 ml. Pass this solution through an ion-exchange
column (internal diameter 2.3 cm) containing 70 ml of Dowex 1×8 resin in the
chloride form. Elute with 2*M* hydrochloric acid and collect 300 ml of eluate
containing interfering radionuclides such as ^{66}Ga, ^{67}Ga, ^{44m}Sc, ^{55}Co, ^{56}Co, ^{48}V and
^{52}Mn (fraction 1). Elute with 6*M* nitric acid and collect 300 ml of eluate containing
^{111}In, ^{201}Pb, ^{203}Pb, ^{62}Zn, ^{65}Zn and some ^{66}Ga and ^{67}Ga (fraction 2). Elute with 0.5*M*
sulphuric acid and collect 300 ml of eluate containing ^{205}Bi and ^{206}Bi (fraction 3).
^{201}Pb and ^{203}Pb are separated from fraction 2 by electrochemical anodic deposition.
Since Cl$^-$ is oxidized at the anode at a lower potential than lead, the solution to be
electrolysed must be free from chloride. Fraction 2 is therefore collected in two
150-ml fractions (2a, 2b). Fraction 2a, which contains chloride ions, is evaporated to
dryness and the residue is taken up in fraction 2b. After electrolysis the solution
contains ^{111}In, ^{62}Zn, ^{65}Zn, ^{66}Ga and ^{67}Ga. No further purification of ^{111}In is
necessary for cadmium concentrations above 0.5 μg/g. Fraction 3 contains ^{205}Bi and
^{206}Bi and the latter can be measured as such.

The detailed procedure is as follows. Evaporate fraction 2a to dryness and take
up the residue in fraction 2b. Dilute to obtain 300 ml of 3*M* nitric acid solution and
add 20 mg of lead(II) oxide, 300 mg of copper(II) nitrate and a few drops of

concentrated sulphuric acid. Heat the solution to 70–90°C, with stirring, place the electrodes in the solution and electrolyse for 1.5 hr at a current of 5 A. Dissolve the lead(IV) oxide deposited, with 50 ml of 3M nitric acid with some added sodium nitrite. Transfer the solution to a polyethylene container for counting. Evaporate the solution remaining after the electrolysis to 50 ml and fraction 3 to 150 ml and transfer them to polyethylene containers for counting.

Figure 6.7 shows a γ-ray spectrum of the solution obtained from the lead(IV)

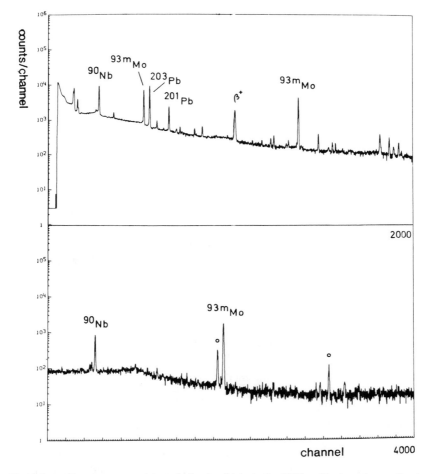

Fig. 6.7 — γ-Ray spectrum of the solution in which the lead(IV) oxide deposit was dissolved. Sample NBS-SRM 1633a; irradiation, 1 hr, 500 nA, 23 MeV protons; decay time, 12 hr; measuring time, 10 hr; ○ — background. (Reproduced by permission, from G. Wauters, C. Vandecasteele, K. Strijckmans and J. Hoste, *J. Radioanal. Nucl. Chem. Articles,* 1987, **112**, 23. Copyright 1987, Akadémiai Kiadó, Budapest.)

oxide deposit. Except for ^{90}Nb, ^{93m}Mo, only ^{203}Pb and ^{201}Pb are detected. Figure 6.8

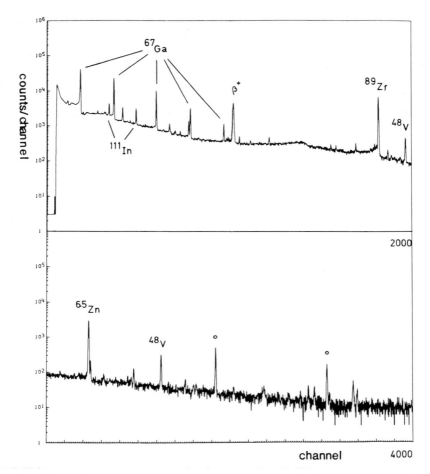

Fig. 6.8 — γ-Ray spectrum of fraction 2 after electrolysis. Sample, NBS-SRM 1633a; irradiation, 1 hr, 500 nA, 23 MeV protons; decay time, 5 d; measuring time, 15 hr; ○ — background. (Reproduced by permission, from G. Wauters, C. Vandecasteele, K. Strijckmans and J. Hoste, *J. Radioanal. Nucl. Chem. Articles,* 1987, **112**, 23. Copyright 1987, Akadémiai Kiadó, Budapest.)

shows a γ-ray spectrum of fraction 2 after electrolysis. The ^{111}In peaks are clearly visible in the spectrum. Fraction 3, which contains ^{206}Bi, is radiochemically pure. The yield of the chemical separation was determined by tracer experiments and was between 99.5 and 100% for indium, lead and bismuth. The irradiated cadmium oxide standard was dissolved in 50 ml of 3M nitric acid, and the thallium oxide and lead oxide were dissolved in 50 and 150 ml of 6M hydrochloric acid, respectively. The samples and standards were measured with a Ge(Li) detector, in the same geometrical conditions.

The detection limits are 6, 44 and 25 ng/g for cadmium, thallium and lead, respectively, for a 1 hr irradiation with a 1 μA beam of 23 MeV protons, followed by a 15 hr measurement after a decay time of 2 days for cadmium and lead and a decay time of 10 hr for thallium. It is assumed that ^{111}In, ^{201}Pb and ^{206}Bi are measured as a precipitate and as close as possible to the detector. When solutions of 50 ml are measured as in the actual procedure described, the detection limits increase by a factor of about 3.

The interference factors (IF) for tin and indium in the determination of cadmium were determined experimentally and are given in Table 6.8, along with the detection

Table 6.8 — Detection limits for cadmium and interference factors (IF) for tin and indium in the determination of cadmium. (Reproduced from G. Wauters, C. Vandecasteele, K. Casteleyn and K. Strijckmans, *J. Trace Microprobe Tech.*, 1987, **5**, 169, by courtesy of Marcel Dekker, Inc.)

Energy, MeV	L_D, ng/g	IF_{Sn}	s	IF_{In}	s
23	6	1.74×10^{-2}	0.03×10^{-2}	0.094×10^{-2}	0.003×10^{-2}
21	10	1.39×10^{-2}	0.02×10^{-2}	0.044×10^{-2}	0.001×10^{-2}
18	13	0.59×10^{-2}	0.01×10^{-2}	0.018×10^{-2}	0.001×10^{-2}
15	34	$< 0.02 \times 10^{-2}$			

s: standard deviation calculated from counting statistics.

limits for cadmium. The nuclear interference from indium is negligible for all proton energies, but that from tin is more important. Since in many environmental samples the tin concentration is higher than the cadmium concentration, a correction is necessary when proton energies above 15 MeV are used. When the tin concentration is known, the cadmium concentration is given by Eq. 6.3 (see Section 2.4):

$$c_{Cd} = c'_{Cd} - IF c_{Sn} \tag{6.3}$$

The nuclear interference from tin can be avoided by using a proton energy of 15 MeV. At this energy the IF is less than 2×10^{-4} and the detection limit is 34 ng/g. Table 6.9 summarizes the results. For cadmium, irradiations were carried out

Table 6.9 — Results for cadmium, thallium and lead in environmental samples. (Reproduced from G. Wauters, C. Vandecasteele, K. Casteleyn and K. Strijckmans, *J. Trace Microprobe Tech.*, 1987, **5**, 169, by courtesy of Marcel Dekker, Inc.)

		$\bar{x}$	s	n	Certified value ± uncertainty
NBS-SRM 1633a					
Cd (μg/g)	15 MeV	0.890	0.030	3	1.00 ± 0.15
	23 MeV	0.943	0.056	4	
Tl (μg/g)	23 MeV	5.73	0.28	3	5.7 ± 0.2
Pb (μg/g)	23 MeV	73.8	1.2	4	72.4 ± 0.4
BCR-CRM 176					
Cd (μg/g)	12 MeV*	477	11	3	470 ± 9
	15 MeV	474	13	3	
	23 MeV	485	20	4	
Tl (μg/g)	23 MeV	2.72	0.11	4	2.85 ± 0.19
Pb (mg/g)	23 MeV	11.04	0.26	4	10.87 ± 0.17
	23 MeV*	11.20	0.40	4	

* Instrumental determination.

with 23 and 15 MeV protons. The results obtained with 23 MeV protons were corrected for the interference from tin and are in good agreement with those obtained with 15 MeV protons. For BCR-CRM 176, the results of instrumental determinations of cadmium and lead, with 12 and 23 MeV protons, respectively, are also given, and are in excellent agreement with the other results. For all three elements and for the two materials, the results are in good agreement with the certified values. The precision ranges from 1.6 to 6%. Additional results obtained by the method described are given in [4].

6.3.4   Analysis of biological samples
Although determination of trace and ultra-trace elements in biological samples is of much interest, only a few attempts have been made [2, 6] to analyse biological samples by CPAA. The reason is obviously that problems arising from sample volatilization or destruction due to heating under irradiation, are still more severe than for most environmental samples. In our opinion, the solution is to ash (high or low temperature ashing, depending on the volatility of the elements determined) the samples before irradiation and analyse the ash. At the same time, this results in improved detection limits, because of the preconcentration.

REFERENCES
[1] A. C. Xenoulis, A. E. Aravantinos and C. E. Douka, *J. Radioanal. Chem.*, 1983, **77**, 207.
[2] L. Zikovsky and E. A. Schweikert, *J. Radioanal. Chem.*, 1977, **37**, 571.
[3] G. Wauters, C. Vandecasteele and J. Hoste, *J. Radioanal. Nucl. Chem., Articles*, 1986, **98**, 345.
[4] G. Wauters, C. Vandecasteele, K. Casteleyn and K. Strijckmans, *J. Trace Microprobe Tech.*, 1987, **5**, 169.
[5] S. Landsberger, A. Giovagnoli, J. L. Debrun and P. Albert, *Int. J. Environ. Anal. Chem.*, 1984, **16**, 295.
[6] M. Bonardi, C. Birattari, M. C. Girardi, R. Pietra and E. Sabbioni, *J. Radioanal. Chem.*, 1982, **70**, 337.
[7] R. Delmas, J. N. Barrandon and J. L. Debrun, *Analusis*, 1976, **4**, 339.
[8] G. Desaedeleer and C. Ronneau, *J. Radioanal. Chem.*, 1976, **32**, 117.
[9] G. Desaedeleer, C. Ronneau and D. Apers, *Anal. Chem.*, 1976, **48**, 572.
[10] K. Strijckmans, N. De Brucker, C. Vandecasteele, *J. Radioanal. Nucl. Chem. Lett.*, 1985, **96**, 389.
[11] G. Wauters, C. Vandecasteele and J. Hoste, *J. Radioanal. Nucl. Chem., Articles*, 1987, **110**, 477.
[12] G. Wauters, C. Vandecasteele, K. Strijckmans and J. Hoste, *J. Radioanal. Nucl. Chem., Articles*, 1987, **112**, 23.
[13] P. Priest, M. Devillers and G. Desaedeleer, in *Radiochemical Analysis*, W. S. Lyon (ed.), p. 191. Ann Arbor Science, Ann Arbor, Mich., 1980.
[14] B. Parsa and S. S. Markowitz, *Anal. Chem.*, 1974, **46**, 186.

7

Activation analysis with heavy ions

Heavy ions ($Z \geqslant 3$) of sufficient energy may induce a large number of nuclear reactions for a given target nuclide. The multiple radioactivation pathways and products increase the prospects for a procedure of high sensitivity, but can also limit its selectivity, as there are more possibilities of producing the same radionuclide from different elements. When, however, the particle energy is limited to below 1 MeV per nucleon, the Coulomb barrier limits activation to nuclides with $Z \leqslant 8$ [1]. In this way, light elements can be determined without interference from the matrix and, in general, with no need for chemical separations.

Barros Leite and Schweikert [2] have given a survey of heavy-ion induced reactions of light elements, with the threshold energies and Coulomb threshold energies. The possibilities of activation analysis with N, Li, B, Be and C beams yielding radionuclides with half-lives of $10-10^4$ sec (^{19}Ne, ^{17}F, ^{14}O, ^{15}O, ^{13}N and ^{18}F) from low Z elements (from hydrogen to fluorine) were considered. The information given [2] was obtained by using a simple computer code and allows selection of the appropriate activation reaction for the analytical problem of interest.

Table 7.1 gives a literature survey of the applications of heavy-ion activation analysis.

Although activation analysis with heavy ions is in principle similar to activation analysis with light charged particles, its application to bulk analysis is in general more difficult, because:

(1) the ranges are very small (typically a few mg/cm^2 for particles with an energy of 1 MeV per nucleon), so inhomogeneity plays a more important role, and, for elements such as carbon, nitrogen and oxygen, the influence of the surface is more important;

(2) the range of the recoil nuclei is, in general, not negligible compared to that of the incident particle; chemical etching after the irradiation can therefore not always be used to remove the influence of the surface;

(3) the ranges and stopping powers for heavy charged particles are less well known [11] than those for light charged particles, so standardization is less accurate.

Table 7.1 — Literature survey of the applications of heavy ion activation analysis.

Reference	Element determined	Projectile	Reaction	Matrix analysed	Concentration range, $\mu g/g$
[3] McGinley *et al.*	H	^{10}B, 40 MeV	$^1H(^{10}B,\alpha)^7Be$	Ti	30–100
[4] Friedli *et al.*	H	^{18}O, 125 MeV	$^1H(^{18}O,n)^{18}F$		
	S		$S(^{18}O,x)^{47}V$		
	Si		$Si(^{18}O,x)^{43,44}Sc$		
	B		$B(^{18}O,x)^{27}Mg$		
[5] Lass *et al.*	Li	^{14}N, 12.5 MeV	$^6Li(^{14}N,\alpha)^{18}F$	glass	60000
			$^7Li(^{14}N,t)^{18}F$		
			$^9Be(^{14}N,\alpha n)^{18}F$	glass	30–500
				rock	1–20
[6] Lass *et al.*	Li	^6Li, 7 MeV			
	Be				
	B				
	C				
	N				
	O				
[7] Ojo *et al.*	C	^6Li, 7 MeV	$^{12}C(^6Li,\alpha n)^{13}N$	steel	50–50000
[1] Lass *et al.*	C	^6Li, 17.5 MeV	$^{12}C(^6Li,\alpha n)^{13}N$	steel	70–6000
	N	^9Be, 13.5 MeV	$^{14}N(^9Be,\alpha n)^{18}F$	Mg	5–11
	B		$^{10,11}B(^9Be,xn)^{18}F$	Si	6–9
	Li	^{12}C, 12 MeV	$^7Li(^{12}C,n)^{18}F$	glass	40–400
				rock	100
	Be	^{14}N, 12.5 MeV	$^9Be(^{14}N,\alpha n)^{18}F$	glass	30–500
[8] Friedli *et al.*	H	^7Li, 21 MeV	$^1H(^7Li,n)^7Be$	Ti	100
				bronze	100
[9] Friedli *et al.*	H	^{10}B, 27 MeV	$^1H(^{10}B,\alpha)^7Be$	Ti	100
				bronze	50
[10] Friedli *et al.*	H	^7Li, 21 MeV	$^1H(^7Li,n)^7Be$	bronze	30
		^{10}B, 27 MeV	$^1H(^{10}B,\alpha)^7Be$	iron	10–40

This last problem can be overcome by using the 'two reaction method' developed by Ishii *et al.* [12]. Two reactions must be selected that yield detectable activities from the analyte element A and a matrix element B, in the sample (X) and in the standard (S). If A and B give suitable reactions with the particle of interest at the same incident energy (E_I), it can be shown that [12]:

$$\frac{a_S^A(E_I)a_X^B(E_I)}{a_S^B(E_I)a_X^A(E_I)} = \frac{n_S^A n_X^B}{n_S^B n_X^A} \tag{7.1}$$

The superscripts A and B refer to the elements A and B, the subscripts S and X to the sample and the standard. The other symbols have the same meaning as in Eqs. (2.47) and (2.50). Equation (7.1) allows the concentration of the element A in a sample containing element B in known concentration (for instance a matrix element) to be deduced from the activities produced from A and B in the sample and in a standard that contains a known amount of A and B. No stopping power data are needed. The validity of Eq. (7.1) is based on the same assumptions as for the average stopping power method (Section 2.3.6). Boulton and Ewan [13] proposed a similar equation.

In addition, the practical detection limits for activation analysis with heavy ions are rather high, e.g. 2 μg/g for the determination of lithium and beryllium in rocks and glass [5], and 5 μg/g for the determination of carbon in steel [7]. For this reason, and because of the difficulties mentioned earlier, use of heavy ions in activation analysis can only be recommended in cases where activation analysis with light projectiles cannot be used.

An example is the determination of hydrogen in metals, for which there is great industrial interest. Hydrogen in metals is usually determined by a hot extraction method. In order to control the accuracy of this method a second, independent, analytical technique is required. Measurement of the prompt γ-rays from the ^{1}H(^{15}N, $\alpha\gamma$)^{12}C reaction allows in principle a bulk hydrogen determination, but the detection limit is rather high. The method is more suitable for profiling the hydrogen concentration as a function of depth. McGinley et al. [3] and Friedli et al. [8–10] used heavy-ion activation analysis through 'inverse reactions'. The reactions used were the ^{1}H(^{7}Li, n)^{7}Be reaction, the inverse of the ^{7}Li(p, n)^{7}Be reaction, and ^{1}H(^{10}B, α)^{7}Be, the inverse of the ^{10}B(p, α)^{7}Be reaction. The radionuclide produced, ^{7}Be, can easily be measured because of its long half-life (53.3 d) and its specific γ-ray energy (477 keV). For 2 hr irradiation at a beam intensity of 25 nA with 21 MeV ^{7}Li, the detection limit is 2 μg/g and nuclear interferences are to be expected only from boron and magnesium [10]. For a 2 hr irradiation at a beam intensity of 320 nA with 27 MeV ^{10}B, the detection limit is 0.5 μg/g, also with possible nuclear interferences from boron and magnesium [10]. The method was applied to the determination of hydrogen in titanium [3, 8, 9], lead bronze [8–10], and iron [10]. For titanium the results were in good agreement with the certified value; for lead bronze and iron both methods yielded comparable results.

A problem with the determination of hydrogen by heavy-ion activation analysis is that the hydrogen concentration is much higher at the surface of metals than in the bulk. McGinley et al. [3] state that one of the advantages of their method is that the influence of hydrogen at the surface can be overcome by etching the sample after the irradiation. This is indeed the case for activation with light charged particles, because the range of the projectiles is much larger than that of the heavy product nuclides. For nuclear reactions of ^{7}Li and ^{10}B with ^{1}H, the recoil range of the ^{7}Be is of the same order of magnitude as that of the ^{7}Li projectiles. In order to demonstrate this experimentally [14], a kapton foil (a polymer that contains hydrogen) with a stack of aluminium foils behind it, was irradiated with 30 MeV ^{7}Li. Figure 7.1 shows the results. The activity in each foil is given as a function of the total depth (kapton foil + stack) expressed in mg/cm^2. No activity was detected in the kapton foil: all the activity was in the fourth to ninth aluminium foils. By kinematic calculations and using the tables of Northcliffe and Schilling [11] that give the ranges for heavy ions, it can be deduced that ^{7}Be formed at the front of the kapton foil should recoil into region 1, and ^{7}Be formed at the back into region 2, in good agreement with the experimental result. The effective range of 30 MeV ^{7}Li is also indicated: ^{7}Be from the kapton is found over the whole ^{7}Li range. At 30 MeV, the influence of surface hydrogen can thus not be avoided by chemical etching after irradiation. The experimental work described indicates that the average range for ^{7}Be is ca. 11 mg/cm^2. This is in reasonable agreement with the 10 mg/cm^2 obtained by Friedli et al. [8] for 35 MeV ^{7}Li by comparing the positions of the cross-section maxima with those deduced from the inverse reaction. Friedli et al. [8] claim, however, that for 21

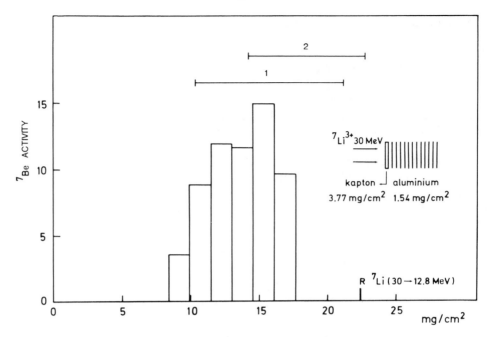

Fig. 7.1 — Activity as a function of the total depth (mg/cm²) in a stack of aluminium foils irradiated behind a kapton foil, with 30 MeV ⁷Li [14].

MeV ⁷Li the recoil range of ⁷Be is only ca. 1.7 mg/cm², so to avoid bulk contamination completely, the samples can be chemically etched after irradiation.

REFERENCES

[1] B. D. Lass, N. G. Roché, A. O. Sanni, E. A. Schweikert and J. F. Ojo, *J. Radioanal. Chem.*, 1982, **70**, 251.
[2] C. V. Barros Leite and E. A. Schweikert, *J. Radioanal. Chem.*, 1979, **53**, 173.
[3] J. R. McGinley, L. Zikovsky and E. A. Schweikert, *J. Radioanal. Chem.*, 1977, **37**, 275.
[4] C. Friedli, B. D. Lass and E. A. Schweikert, *J. Radioanal. Chem.*, 1979, **54**, 281.
[5] B. D. Lass, C. Friedli and E. A. Schweikert, *J. Radioanal. Chem.*, 1980, **57**, 481.
[6] B. D. Lass, J. F. Ojo and E. A. Schweikert, *J. Radioanal. Chem.*, 1980, **60**, 255.
[7] J. F. Ojo, B. D. Lass and E. A. Schweikert, *J. Radioanal. Chem.*, 1980, **60**, 261.
[8] C. Friedli, E. A. Schweikert and P. Lerch, *J. Radioanal. Nucl. Chem.*, *Articles*, 1985, **88**, 369.
[9] C. Friedli, E. A. Schweikert and P. Lerch, *J. Radioanal. Nucl. Chem.*, *Articles*, 1985, **90**, 341.
[10] C. Friedli, E. A. Schweikert and P. Lerch, *J. Radioanal. Nucl. Chem.*, *Articles*, 1985, **90**, 349.
[11] L. C. Northcliffe and R. F. Schilling, *Nuclear Data Tables*, 1970, **A7**, 233.
[12] K. Ishii, C. S. Sastri, M. Valladon, B. Borderie and J. L. Debrun, *Nucl. Inst. Meth.*, 1978, **153**, 507.
[13] R. B. Boulton and G. T. Ewan, *Anal. Chem.*, 1977, **49**, 1297.
[14] C. Vandecasteele, Unpublished results.

Conclusion

It has been shown that activation analysis with charged particles is a useful analytical method with the following main areas of application:

(1) determination of light elements, mainly in metals and semiconductors;
(2) determination of medium and heavy elements in metals and semiconductors;
(3) analysis of geological and environmental samples.

The method is very valuable as a complement to neutron activation analysis: it allows the sensitive determination of many elements that cannot be determined by neutron activation analysis and also allows the instrumental analysis of some matrices that, for neutron activation analysis, require radiochemical separations.

When particular attention is paid to the choice of standards, and an accurate method for standardization is used, CPAA is a very accurate method. A particular advantage is that a surface layer can be removed from the sample after the irradiation, in order to remove surface contamination. The accuracy has been demonstrated by various interlaboratory and intermethod comparisons. CPAA has also been used frequently and with success to certify reference materials.

Mostly, use has been made of protons, deuterons, helium-3 and helium-4 as projectiles, but more recently heavy ions ($Z \geqslant 3$) have also been used.

Index